EVOLUTION

OTHER WORKS BY JEAN-PIERRE ROGEL

L'hippopotame du Saint-Laurent: dernières nouvelles de l'évolution (2007)

La grande saga des gènes (1999)

Le défi de l'immigration (1989)

Bombe A à Montréal, documentary film (1989)

Un paradis de la pollution (1981)

Face au nucléaire, co-authored (1979)

EVOLUTION

THE VIEW FROM THE COTTAGE

JEAN-PIERRE ROGEL

Translated by Nigel Spencer

RONSDALE PRESS

First published as *L'hippopotame du Saint-Laurent* (Multimondes, 2007)

RONSDALE PRESS
3350 West 21st Avenue, Vancouver, B.C.
Canada V6S 1G7
www.ronsdalepress.com

Typesetting: Julie Cochrane, in Granjon 11.5 pt on 15
Cover Design: Cyanotype
Cover Photo Credit: Ivan Stanic
Index: Noah Moscovitch
Paper: Ancient Forest Friendly "Silva" (FSC) — 100% post-consumer waste, totally chlorine-free and acid-free

Ronsdale Press wishes to thank the following for their support of its publishing program: the Canada Council for the Arts, the Government of Canada through the Canada Book Fund, the British Columbia Arts Council, and the Province of British Columbia through the British Columbia Book Publishing Tax Credit program.

Ronsdale Press also acknowledges the financial support of the Government of Canada, through the National Translation Program for Book Publishing, for our translation activities.

Library and Archives Canada Cataloguing in Publication

Rogel, Jean-Pierre, 1950–
Evolution: the view from the cottage / Jean-Pierre Rogel; translator, Nigel Spencer.

Translation of: L'hippopotame du Saint-Laurent.
Includes bibliographical references and index.
ISBN 978-1-55380-104-7

1. Evolution (Biology). 2. Biodiversity. 3. Nature — Effect of human beings on. 4. Nature conservation. I. Title.

QH367.R6313 2010 576.8 C2010-904858-X

At Ronsdale Press we are committed to protecting the environment. To this end we are working with Canopy (formerly Markets Initiative) and printers to phase out our use of paper produced from ancient forests. This book is one step towards that goal.

Printed in Canada by Marquis Printing, Quebec

Acknowledgements

I would first like to thank the readers of my column *Planète ADN* in the magazine *Québec Science*: this book owes much to their support over the years. I would also like to recognize the contribution of Joël Leblanc, who researched three chapters and wrote up material which provided some very stimulating starting points. His paleontological training helped me find my way through the arcane world of fossils. Chapters 7 and 13 bear his imprint. I wish also to thank Robert Loiselle, professor of evolutionary science at the Université du Québec at Chicoutimi. I much appreciate his attentive reading of the manuscript and his generous, constructive remarks. The great evolutionary specialist Cyrille Barrette of Université Laval is owed great thanks for having read the manuscript and made invaluable suggestions. Pierre Béland, specialist in marine mammals, has kindly examined Chapter 4. Thanks are due to my colleagues Claude D'Astous, Hélène Courchesne and Michel Rochon for their advice on an earlier version. Any error or inaccuracy, however, remains my responsibility.

For the preparation of this book, the Centre du livre français provided support in the form of a grant, for which I thank them.

This English edition is based on the original version in French, but has been enlarged and updated to take into account recent developments in science. I want to thank Ronald Hatch of Ronsdale Press for his guidance through the process, and — last but not least — Nigel Spencer in his double role of attentive friend and impeccable translator.

CONTENTS

Introduction / 1

PART ONE:
Connected by
an Invisible Thread

"Mystery of Mysteries":
The View from the Cottage / 9

The Tree of Life and the Two "D"s / 21

And What's Your Bar Code?/ 32

Hippos in the St. Lawrence / 39

PART TWO
All Parts Included:
Some Assembly Required

The Fly and the Butterfly Tell All / 55

Of Finches and Their Beaks / 65

Stuck in the Mud and How to Get Out
in under 10 Million Years / 74

The Thumb and the Baby Panda / 83

A Very Brainy Animal / 89

PART THREE
When Humans Interfere with Evolution

Parasites in High Gear / 101

Salmon under the Influence / 108

Mr. O'Connor's Stubborn Struggle / 116

Avatars of the White Bear / 125

Sign of the Loon / 133

Calendar of life / 145

Glossary / 147

Notes / 153

Bibliography / 157

About the Author & Translator / 161

Index / 163

Introduction

JUST OVER 150 YEARS AGO, in June 1858, the English naturalist Charles Darwin, tucked away in the Kent countryside, received a manuscript from a young colleague, Alfred Russell Wallace, presenting his thoughts on nature. So innovative were these ideas, and so like Darwin's own, finally about to be published after twenty years, that he determined to work on a joint publication with Wallace for the Linnean Society. This marked the beginning of a revolution in science, and a year later, Darwin published *On the Origin of Species*. The first printing sold out in a single day — lucky author! — and it is safe to say that nothing in biology was ever the same again.

Their main idea that species evolve and descend from one another — most of them disappearing in the great expanse of time — would change the way we see the world. If Darwin and Wallace had merely been content to present this theory, already advanced by others, their influence would not have been as great. But they went further, explaining natural selection, the mechanism by which evolution occurred, and convincingly showing how it functioned. Because religious dogmas at the time preached that species were fixed and unchanging, an inevitable clash occurred.

Today, having just celebrated the 150th anniversary of its publication with exhibitions, books and films, the theory of evolution by natural selection is, for scientists, unshakeable. Of course, it does not explain everything in complete detail, even in much-studied organisms like mice, even more so in humans, but it is a solid scientific theory, tested and proven, despite repeated criticisms for more than a century. It is the indispensible framework in which to explain life. As the American geneticist Theodosius Dobzhansky said in 1973, "Nothing in biology makes sense except in the light of evolution."

During the filming of a documentary for Radio-Canada in 2008, I asked Richard Dawkins of Oxford University — a well-known defender of Darwin, who has written eight books on the biology of evolution — about the role of Darwin's ideas. Here is how he replied:

> I think that Darwin's idea is perhaps the most powerful idea that any human mind ever had in the sense that it did the most explanatory work that actually changed the way people think, because before Darwin came along, the whole of the living world, all this magnificent complexity and beauty and elegance and diversity had no explanation at all. People knew it was there, and they were describing it, but nobody knew what caused it; nobody knew how it came into being. Darwin changed all that.

For the rest of society, it may not be so clear. Among the broad public, the theory's success is mixed. Often poorly known or understood, it is easily confused with gross oversimplification ("survival of the fittest," for example). It is frequently challenged by fundamentalist religious belief, which is experiencing a resurgence across the globe, and my personal crystal ball tells me it will be a hot topic for years to come. Creationists or neo-creationists in the Intelligent Design movement will redouble their attacks on evolution, and this is all the more reason to discuss it publicly and show its full power and subtlety.

This public debate is partly the context that gives rise to the present book. It appears useful nowadays to discuss these fundamental ideas about living things, as well as the gulf that separates scientists from the rest of society. If science is publicly repudiated, it loses credibility and its ability, notably among decision-makers, to solve important problems for the planet, especially climate change and the massive loss in biodiversity. Of course, not everything can be solved by science, but it does provide for an exchange of ideas that esoteric or religious beliefs cannot replace without leaving humanity and the planet at great risk.

This book springs also from the desire to provide concrete examples to show how the science of evolution has been refined over the past 150 years. Today, by integrating modern learning in genetics and molecular biology, that science is more powerful and unassailable than ever. Both Darwin and the ideas that have developed in his wake are important and fascinating.

For this reason, I have chosen a journalistic approach aimed at drawing attention to the newest elements of evolution. The "star," it might be said, is what has come to be called "evo-devo," a contraction used by specialists for "evolution and development." Emerging in the past fifteen years, evo-devo is a new way of approaching evolution that relies on recent discoveries in the biology of embryo development and in comparative genetics. The expression evo-devo may, at the outset, seem a sort of specialist jargon and repel the uninitiated. It does lead, however, to a newer and deeper look at the world of living things, and it is an approach we shall hear more and more about.

In order to deal with the science of evolution, I have presented evo-devo themes and accomplishments through sample case histories. Frequently, a particular anecdote or situation leads to a discussion of a question with far broader implications, and a fresh perspective offered by modern biology. Thus, each chapter can be read as a stand-alone essay, much after the manner of Stephen Jay Gould, a true master in the field. As a consequence, the reader might want to approach them in no particular order and refer to the glossary when encountering words or concepts that are unfamiliar.

There is, nevertheless, a thread and a progression to the ideas, as indicated by their division into three parts. In the first chapter, I offer the familiar example of a wooded area in southern Quebec, revisiting what we have learned from Darwin and connecting it to what modern science has shown us. Then in Chapter 2 we come to what DNA analysis has contributed to Darwin's intuitive but scientifically well-founded concept of a tree of life that includes all species. We will then see (Chapter 3) how all this knowledge applies to our catalogues of biodiversity and the review of the development of a well-known sea mammal, although perhaps not from an evolutionary viewpoint (Chapter 4).

In the second part we move into the thick of evo-devo and look at recent discoveries in architect genes that govern the making of animals (Chapter 5). The next chapter takes us into the world of finches, above all the famous Galapagos finches discovered by Charles Darwin (Chapter 6), but as we shall see, it is very much a story both contemporary and universal, concerning beaks, genes and climate change. Then we turn to two applications: the creation of paws from fins (Chapter 7) and the panda's curious thumb (Chapter 8). Next comes the sensitive topic of the disturbing genetic proximity between humans and chimpanzees (Chapter 9).

The third and last part deals with how humans play with the machinery of evolution — so much so that evolutionary changes have become rapid enough for scientists to refer to them paradoxically as "contemporary evolution." We shall see examples of these changes but also some additional examples of species conservation, for if we are capable of harming animal and plant species, we can also help in their conservation.

Throughout, I have kept in mind all those who like to involve themselves in nature in their moments of leisure, and to this end I have often described personal experiences at our lakeside cottage. This is nature recomposed, of course, not untouched nature in the wildest state — as if such a thing still existed. Be that as it may, these are areas rich in animal and plant life that we cling to and wish to protect. To protect well, one must know well.

First, then, this book can be seen as an invitation to take a fresh look at nature as it surrounds us, wherever we are, in town or country. Canadians are certainly privileged to have ready access still to large swaths of nature, even wild nature reserves. Although many visit them, they may be unaware of what is offered there. No matter where we live in this country, we must become aware of the importance of the riches around us, riches we many not even suspect to be there.

Second, it has been my goal to lead readers toward science itself and show the strength of what it does at an essential level that concerns us all. It is not the simplest thing to explain how life forms are built, to explain the source of biodiversity, or what sets humans apart from the primates. In the background there is always the sense of how this touches us personally: each of us different, all of us cousins. The idea that all living things — from bacteria to men, salmon to birches — share the same genetic code has enormous implications. Finally, as Stephen Jay Gould says in *The Panda's Thumb*:

> And then, of course, there are all those organisms: more than a million described species, from bacterium to blue whale, with one hell of a lot of beetles in between — each with its own beauty, and each with a story to tell.[1]

The following pages contain a few of these stories drawn from recent research, often somewhat technical, though I have tried to tidy up the technical jargon. My goal is, above all, to offer the broadest possible public an essential part of what modern science has to offer, something I believe each of us can benefit from: an appreciation of the basis of living things.

PART ONE

Connected by an Invisible Thread

"Mystery of Mysteries": The View from the Cottage

AS WE MAKE OUR way down the pine-covered path loaded with supplies, a prolonged cry echoes through the woods, a sort of "chir-t-t-t" sound, sharp and insistent, amazingly loud for where it's coming from: a 200-gram ball of fur sitting on a pine branch. It's Friday evening, we've come up to the cottage, and Oscar the red squirrel is there with his usual loud welcome.

You have to understand Oscar. This bit of forest in southern Quebec is his home, and we're a nuisance. All weekend long, he'll remind us of this by letting out his cry and scurrying frantically along the gutters at the edge of the roof. It's the end of May, beautifully warm and sunny weather, so we'll be eating all our meals outside, but Oscar couldn't care less. He'll still tear along the edge of the roof on his little paws, just a metre above us. Then in one bound, he'll leap into the cedar at the corner of the cottage and be off on his highway of interlaced trunks and branches.

This is how life goes on at the foot of the elephant mountain, a modest hill in the Eastern Townships of southern Quebec. Officially, it rises 525 metres, just a hillock on the western shore of Lake Memphremagog, one hundred kilometres west of Montreal. It's too humble to claim the title "Mount Elephant," so I'll just follow tradition and call it the elephant mountain.

Although this protuberance is beautifully forested, it's no place for a hike; there's nothing you could call a trail, up or even around it. It does provide us with a cosy neighbour and a reference point on the horizon facing our lakeside cottage. In the morning, we stare at this pachyderm. Any clouds over the elephant mountain? Hmm, bad sign if the sun's going in. The elephant's head and trunk stretching off to the right are silhouetted by blue? That's a good sign; it means the wind's southwesterly. Apart from that, we really just pay attention to what's going on closer to home. The lake isn't large, but it is in the middle of a forest of leafy trees and conifers which, believe it or not, is still pretty much the same as when the first cottages appeared in the 1960s. The shore is still very wooded and protected by owners' association rules: no outboard motors, for instance. All the cottagers live encased in their own little bit of greenery. Sometimes it's noisy right by the edge of the lake. Not everyone's figured out that voices carry perfectly all the way across. In summer, pedal boats, kayaks and canoes weave their way over the water, and life flows peaceably on.

It is both simple and amazing. We acquired the cottage a few years ago, and I was won over far more than I ever expected by the nature surrounding it. Still, what could be more ordinary than this wooded area and its human modifications? It's not even an ancient forest, just the restored version of a wood that must have been exploited right from the early days of colonization.

The ecological rundown is easy to do: a small, shallow, elongated lake with a catchment area at its head, shores populated by conifers and leafy trees, the usual underbrush, a few mammals here and there (foxes, deer, and above all squirrels and small rodents), birds, numerous small invertebrates, thick mulch, and with all that, the cycle of seasons: a chunk of life, brilliant in its unity and its diversity — really simple and surprising all at once. It's easy to understand, if only by intuition, how these organisms are interconnected and depend on their surroundings, though obviously disturbed by humans.

Still, at a deeper level, how and why have all these things come to be here? What makes them change, renew themselves and disappear? As the English philosopher John Herschel put it, it's the "mystery of mysteries" within easy reach: mystery seen from the cottage.

Today, *Tamiasciurus hudsonicus*, Oscar in this case, is even more excited than usual. He stops at the big cedar, looking toward the lake and chattering like a magpie, his tail twitching in agitation. We've learned the hard way not to trust the little hypocrite. No matter how much he runs around all day collecting seeds, he's not averse to stealing a few chips and even tortillas. He ought to be happy with what he's got already, a nice place full of conifers just right for him. He's got giant white pine rising forty metres, white cedar, balsam fir, hemlock, beech, maple and birch. This "seed-nut" has an embarrassment of riches, and it's amazing to see him plow his way through his favourite, pine cones. In autumn, he takes an occasional break from shaking them out of a tree and hiding them in a myriad of secret places to eat one right under our noses. Ten seconds is all he needs to disassemble and devour the gummy things as though they were ordinary corn on the cob. What a stomach! He's equally opportunistic with buds, flowers and mushrooms.

It's been a calm night, and the leaves are barely rustling, but at 4:30 in the morning, a small group of birds shows up. Half awake, I can just make out five distinct types of song, but that's as far as it goes. I'm not good at identifying them by ear, or by sight for that matter.

I know we have black-headed chickadees year-round, blue jays and American robins, and for the time being, other types of chickadees, as well as warblers — migratory birds, these. The other day I spotted some that were partly yellow, but that's about as much as I know about them. Well, no, there is one bird I really like and know something about: the hairy woodpecker, and he stays year-round. We can hear him drumming away on a large dead tree behind the house. Standing guard, he cleans out all the dead trees and keeps ants and other insects away from the healthy ones nearby. If you come up on him gradually, he won't fly away. I've seen him a few times, and you can tell he's a male by the red marking on the back of his head. With his white belly and black-and-white-striped wings, he looks like a comedian or an opera singer, except that he doesn't sing. His sharp-clawed feet fascinate me. Fingers modified for a better grip — two in the front and two in the back — are complemented by a tail that serves as an additional point of support. Solidly planted in the bark, these claws allow him to hang vertically from a tree trunk and hammer away so fast with his powerful beak that all I can see is a blur. I have read somewhere that his skull is especially well-suited to this strange occupation and it allows him to knock on wood more than a hundred times a minute. Not only is the skull thick, but it is also powerfully muscled. In addition, his beak and bones are not fused as in other birds, but connected with a spongy tissue that serves as a shock absorber.

Oh, how proud we are of *our loons*! The possessive adjective is entirely misleading of course, but our lake is visited by a pair of loony divers. I do love this magnificently plumed fish eater. Since 1987, the loon's silhouette has adorned the Canadian dollar, popularly called "the loonie," even in financial circles. Morning and evening we can hear his call, a very moving tremolo which has come to symbolize the natural beauty of our waters. Our lake's a modest-sized one, so they don't actually nest here, but only stop off for a meal from time to time. The minute we spot them, it's on with the binoculars, and we get as worked up as brokers on the stock-exchange floor. Often they come as a pair and land in the middle of the lake to do some quiet fishing. They swim half-submerged for a minute or two, then suddenly dive. Up they come a way off, and although they can wander quite some distance apart, they're always back together.

It's quite easy to make them out. Their heads and necks are black, with short vertical lines that form a white necklace. The back is a magnificent checkerboard of black-and-white plumage. They take off by running across the water, and their white breast seems to swell as they keep on flapping their wings. Their wing span is an impressive 1.3 metres, but their take-off is heavy, and from my window they resemble bombers or seaplanes rumbling down a runway barely long enough for them, but this doesn't do justice to the animal grace with which they beat their long wings.

Their cry, of course, is notorious. Or is it a song? This aquatic bird, for that's what he is, has a widely varied vocal repertory. While in flight, he lets out a *kwook* sound rather like a barking dog, and at a short distance from one another, adults will ululate softly. But on a summer evening, one cannot help being struck by their really loud cries: the three-note form, then a plaintive howl like a wolf, and finally, a long, sharp cry like hysterical laughter. This range of cries is heard mostly during the mating season, and anyone hearing them at evening, at night or in early morning out on a lake, is enveloped in the deep melancholy of their peculiar sound.

After listening to the birds for a time, I fall asleep again and reawaken around 8:00 a.m. The sun is emerging above the trees across the lake, and I take a few steps into the woods along a small path we've cleared to admire the plants in the underbrush, which are magnificent at this time of year. Then suddenly a mist of light rain seems to come out of nowhere, while the sun goes on shining.

The air is instantly laden with the smell of wet leaves, and I see a thin film fall onto our veranda. It's a mix of pollen and encased seeds from the large pines around us. The rain has threshed out these particles filled with the promise of new life, for they contain DNA, deoxyribonucleic acid. DNA is literally raining from the sky with new instructions for the formation of life: a rain of living, millennial information.

I look up at the white pines, the jewels of our little domain. I've figured out they must be about thirty-five metres high and 120 years old. These trees have cones which work the same way flowers do, and these are the days when their male seed (pollen) is being released into the air. The season is too far on for this pollen to come from the flowering trees nearby, so what's landing on my balcony must be from the pines. It is released and flows gently down from the base of higher-up branches just below the growing female cones. Whatever happens after that is entirely up to the wind. While I was in the Austrian woods some years ago, I noticed one evening that the intense output of pollen creates small clouds that are visible to the naked eye. I have never seen anything quite like it here — not enough pines for that. I think I noticed this spring's output only because of today's sudden rain.

How far would this pollen travel normally? I hope it at least reaches another stand of white pine, the ones scattered around the edge of the lake in any case, so it can land on and fertilize the ovules of female cones. They need sex for reproduction just as much as we do. Self-fertilization is not an option; there has to be a partner of the same species.

Reproduction is one of the mysteries of life, always fascinating to watch, since it connects with something very intimate inside us. It takes some training to spot it in plants and insects, but if you have the patience, it soon becomes obvious, just as it does in the animals that look more like us. In fact, we humans seem to follow a universal pattern of behaviour: we are captivated and moved by the small offspring of our pets or farm animals, and equally touched by a female mammal protecting her young in natural surroundings.

The mystery of individual members in a species succeeding one another was practically resolved at the beginning of the 19th century before Charles Darwin was even born.

Nineteenth-century naturalists had already described various forms of animal and plant reproduction, building a painstaking inventory of living things on their voyages of discovery. In 1735, Carl Linnaeus of Sweden had devised a universal classification by classes, orders, families, genera and species that stood as the authoritative reference. A species represented a reproductively isolated collection of individuals easily able to cohabit and breed with one another in natural surroundings, but not with other similar organisms.

Still, however, naturalists avoided answering some fundamental questions: how did species *themselves* come into being? Are they immutable, or do they change over time, and if so, why? Where do humans fit into this scheme of things? Every culture in every epoch has tried to explain the origins of species, but in the early 19th century, science was still in its near-infancy and could not.

Long-standing and ancient concepts still clouded the debate. First came *fixism* from the Greeks, who held that the world was made of "essences" and thus unchanging. Then as the monotheistic religions emerged (Judaism, Christianity and Islam), *creationism* took its place. In this approach, God was seen as the creator of all from nothing — at once and forever. Humanity then was seen as a co-creator, or as the ultimate achievement of the creative process. In the Middle Ages, some thinkers tempered creationism with the notion of *spontaneous generation*. It took many centuries for this to be refuted.

It was not until the beginning of the 18th century that *evolution*, the idea that lineal descent connected various life forms, began to gain adherents. In 1809, Jean-Baptiste de Monet, Chevalier de Lamarck, published his *Zoological Philosophy*, a milestone. He affirmed that all species were related and derived from one another, although most had disappeared along the way. He postulated that a "life force" propelled all living things to transform themselves gradually, in increasing complexity. This model was referred to as *transformism*. To support this, Lamarck formulated two laws: "the modification of organs according to needs" (with the famous example of long giraffes' necks that enabled them to graze on trees) and "the heredity of acquired characteristics." He was wrong, completely wrong, on both counts. It was not until Darwin published his *On the Origin of Species* fifty years later that this would be corrected.

Charles Darwin was twenty-two when he joined the *Beagle* as a naturalist for a round-the-world trip that would include South America and Australia. This quiet, well-bred Englishman with a conservative turn of mind was most certainly a creationist at the outset, but when he returned five years later, he had changed his mind, and evolution had gained the upper hand. Historians situate this shift during the first few years after his return to England, and from that time he continued to contemplate what he referred to as "my theory." It nevertheless took him another twenty years to publish it, so disturbed was he by the debate it would stir up and so aware of the need to prove it beyond question. Reading his notebook, however, leaves no room for doubt. In March of 1838, for instance, commenting on a letter from John Herschel to Charles Lyell, he exclaimed, "Herschel refers to the appearance of new species as the 'mystery of mysteries,' and he devotes part of his text to the question. Hurrah!" Having found the answer, of course, he was exultant. He knew that he would at once astonish and shock his contemporaries, making a major contribution to science in the process.

When at last *On the Origin of Species* appeared in 1859, the answer was detailed and crystal clear. One paragraph from the introduction sums it all up. Narrow minds might be tempted to say this was his only scientific idea, but it was genius just the same:

> . . . the Struggle for Existence amongst all organic beings throughout the world, which inevitably follows from their high geometrical powers of increase, will be treated of. This is the doctrine of Malthus, applied to the whole animal and vegetable kingdoms. As many more individuals of each species are born than can possibly survive; and as, consequently, there is a frequently recurring struggle for existence, it follows that any being, if it vary however slightly in any manner profitable to itself, under the complex and sometimes varying conditions of life, will have a better chance of surviving, and thus be naturally selected.[2]

Thus was he proposing a thoroughly evolutionist vision of life forms (though he actually used the word "evolution" sparingly, preferring the phrase "descendance with modification"). Above all, however, he proposed and proved the existence of a mechanism to explain what had already been observed: natural selection.

For the rest, Darwin's demonstration is very solid indeed, despite the fact that on certain points his lack of knowledge led him to errors of interpretation. He anticipates the main points of objection to his thesis and counters them methodically. His development relies cannily on daily realities known to people around him, such as the breeding of domestic animals. In fact, this causes him to spend relatively little time talking about the decisive observations he made on the Galapagos Islands. Instead, he dwells on a domain he knows well: pigeon breeding. In four chapters, he proceeds from artificial selection to his key idea: natural selection.

Struck by the wide variety of pigeons artificially developed by breeders, he notes that, observing them in the wild, a naturalist would class them as distinct species. Inquiring as to how such diversity was possible, he evokes the mechanism of selection. The breeder begins with the pigeons that best suit his purpose — say, those with the most colourful plumage — and has them mate. After a number of generations, he finds himself with specimens that have little to do with the wild originals. Darwin shows how the same phenomenon occurs in nature, although randomly and without apparent purpose, natural selection occurring blindly, whatever its context.

Returning to his main theme, Darwin points out that each individual in a given species usually has many descendants, too many for all to survive, reach maturity and reproduce. From birth then, nature sorts them, and only those best adapted to their surroundings manage to survive, while others die. Generation after generation, the blind game of natural selection results in a population that changes, adapts, with disparities between populations accruing over time. Thus one may observe the emergence of new species. Most often this is the result of geographical isolation or changes in environmental conditions. The key fact is that such phenomena will result in the formation of two distinct populations whose members can no longer interbreed. In this way, living creatures form diverse species across *deep time*, in other words, successive geological ages.

Nevertheless, Darwin found himself up against the question of inherited characteristics. Like most of his contemporaries, Darwin did not know that Gregor Johann Mendel, a monk in far away Moravia, discovered and published the answer in 1866. In his now-famous study on heredity in peas, Mendel demonstrated that inherited characteristics depend on material elements transferred by *gametes* (sex cells) when fecundation occurs, and that these are shared out among progeny according to statistical laws. When Mendel's findings were "rediscovered" and republished around 1907, Thomas Hunt Morgan followed up a few years later with a *chromosome* theory of heredity. With his work on the Drosophila fly, he showed that genes, the determinants of hereditary characteristics, are carried by chromosomes in the cell nucleus, and that each individual receives only half of the parents' genes.

This became the point of departure for an entirely new discipline often accompanied by an overwhelming mathematical apparatus, *population genetics*. This added to "Darwinism" — leading to the expression "Neo-Darwinism" — rather than invalidating it. Then, in the 1930s, a synthesis of new knowledge began to emerge, and the likes of Ronald Fisher showed that natural selection favoured the continuation of certain *alleles* (different forms of a single gene) insofar as they conferred an advantage on the population. A new overall understanding now lay in the offing. This is primarily due to George Simpson (paleontologist), Ernst Mayr (biologist) and Theodosius Dobzhansky (geneticist); from their writings emerged what we call the *synthetic theory of evolution*. In the late 1950s and early '60s, the discovery of the DNA double helix by Francis Crick and James Watson, followed by the genetic code, further built onto this already-solid structure.

Enough about the history of ideas for the moment: suffice it to say that the synthetic theory of evolution complemented and enriched Darwinism. Researchers began to express things on a different level. It was reasonable to assume that, if genetic material was subject to mutations transmitted to future generations, an important source of variability would come into play. Darwinian logic still retains its validity in that, once expressed, mutations are well and truly subject to natural selection.

Both approaches remain valid. One may explain biodiversity from a naturalist's point of view: that is, one may be interested in entire organisms as well as the adaptation and reproductive success of one species or another. One may equally explain it from the perspective of a population geneticist and concentrate on genes and their relative proportions in a given population from generation to generation. These standpoints are not contradictory; they simply relate to two distinct levels of the organization of life.

Cyrille Barrette, evolutionary specialist at Université Laval, illustrates the relationship between these two levels with the metaphor of a relay race:

> The runner is the individual, and the baton the genes: each relay represents a generation, and the entire contest is like a partial history of the lineage. In the relay race, the baton being passed from hand-to-hand is essential, for it alone travels the entire distance. If a single runner drops it, the race is over

> for that particular team (ascendancy). One cannot run or win without the baton, but it cannot travel alone: yet each runner is only its temporary carrier, just as any individual is only the temporary bearer of genes. Yet, without a runner, there is no race.[3]

Barrette explains that from a naturalist's point of view, it is the runners that do the hard work — the individual is stimulating and alive — so the race is what they most like to talk about. Still, the importance of the baton — or the program that inhabits every individual and every species — must be acknowledged. The modern vision of living creatures combines these two dimensions.

It is the first weekend of September: a perfect time to take the canoe for a spin around the lake. We ease into what, with a touch of melodrama, we call "the bay in the depths of time." It's simply a remote bay, a small, shallow bit of water with no human habitation nearby, a dark and quiet place filled with aquatic plants and submerged cedar trunks. Steering carefully, we slide the canoe toward some curious plants on the shoreline. They are purple pitcher plants (*Sarracenia purpurea*).

These magnificent red flowers stand erect, about 30 cm high, but more astonishing is what lies at their base: dark-green trumpet-shaped bells veined with purple that open wide to catch rainwater falling from above, as well as the insects that wander into them. These are the carnivorous plants that Brother Marie-Victorin spoke so much about in his famous book *Flore Laurentienne*:

> The purple sarracenia is the most extraordinary of our flora, and the principal ornament in our bogs, a classic example of carnivorousness in plants. Exploring the inside of this tubular leaf with one's finger, one perceives a layer of downward-pointing hairs that allow easy entry but difficult exit for the unfortunate insects that wander into the tiny receptacle seeking shelter or refreshment. Often the imprisoned insect exhausts itself in vain attempts at escape, almost always drowning in the water contained there. A particular diastasis dissolves the captured bodies thus digested by the plant tissue, perhaps allowing the organic substances to be absorbed directly.[4]

One is immediately struck by the originality of this plant and amazed by its ability to trap, break down and finally digest insects. How can all this be the product of evolution? Although few studies on carnivorousness in plants exist due to the absence of fossils, it is possible to reconstruct some plausible stages in the process.

First of all, there have always been, and still are, numerous plant species which have developed apparently in order to hold water, especially desert epiphytes. Some insects fall into these, drown and decompose — not necessarily from the plant itself, but from bacteria. Many plants, moreover, can absorb nutritive substances very well through their leaves, mineral salts in particular. One can well imagine, under certain environmental circumstances, that substances resulting from the decomposition of insects went from being simply a nutritional supplement for a plant to its main source of nourishment. Quite simply, opportunism has made the plant into a carnivore. Then, again by natural selection, other developments along the same lines also occurred: beautiful colours to attract prey, hairs angled in the right direction, nectar in the bottom of the receptacle, digestive enzymes, and so on. If this model of what can be called "progressive construction" is not obvious to the layman, it definitely will be to the biologist or any experienced observer. Nor does this detract in any way from the mysterious beauty of such a magnificent plant.

The Tree of Life and the Two "D"s

IF YOU HAVE THE option, mid-May is the best time to visit the Jardin botanique de Montréal. It's one of those places where, after our long, rough winter, the intensity of spring is best felt. Barely a month after the last snow, plants have already shaken off their lethargy and are growing at a tremendous rate, as though making up for lost time with a twinge of guilt and urgency. Among the Jardin's seventy-five hectares at the base of the famous Olympic Stadium are virtually all the plant species native to this latitude, as well as many others from around the world.

Every spring when I visit, my first stop is the apple trees and flowering lilacs. I love their floral exuberance and splendid rows of bright colour, although the excessive regularity of the layout can be annoying. After that, I head straight for the undergrowth. There, shaded by large maples, ashes and lindens, grow a number of spring plants, most of all primroses and bloodroot. Until the leaves' shade catches up with them and hides the sun, they are in full riot. The city's noise is barely audible; this might as well be the depths of the forest. There are always birds in the bushes trilling loudly. In a few weeks, the trees will be in thicker leaf, and the spring plants will give way to species less in need of sunlight — ferns and hostas for example — but knowing the place in summer, it will still be intimate and silent.

The last time I went there, I spent a while in the arboretum looking for a particular tree. The previous summer, I had reread a section of *On the Origin of Species* that caught my attention, one in which Darwin compares all living things to a tree. The image was meant to show the links and kinship that connect all life forms. The quote is a long one, but it merits special attention since it is at the heart of Darwinian thought. It begins like this:

> The affinities of all the beings of the same class have sometimes been represented by a great tree. I believe this simile largely speaks the truth. The green and budding twigs may represent existing species; and those produced during each former year may represent the long succession of extinct species. At each period of growth all the growing twigs have tried to branch out on all sides, and to overtop and kill the surrounding twigs and branches, in the same manner as species and groups of species have ever tried to overmaster other species in the great battle for life.[5]

What emerges is the notion that each species succeeds another and that only the most able individuals survive. This goes to the heart of Darwin's central affirmation, but the English biologist did not explain what it is that links our observations to the geological past, a highly controversial matter in his day. What follows is a slightly abbreviated version of his thoughts:

> . . . of the many twigs which flourished when the tree was a mere bush, only two or three, now grown into great branches, yet survive and bear all the other branches, so with the species which lived during long-past geological periods, a very few now have living and modified descendants. From the first growth of the tree, many a limb and branch has decayed and dropped off. . . .[6]

With the same analogy, he thus affirmed that most species probably died out, leaving either fossilized traces or none at all, and today what we see are the modified descendants of those past species. But he goes on, having to deal with some strange and insular animals, to offer a valid explanation for oddities in nature. Here it is:

As we here and there see a thin straggling branch springing from a fork low down in a tree, and which by some chance has been favoured and is still alive on its summit, so we occasionally see an animal like the Ornithorynchus or Lepidosiren, which in some small degree connects by its affinities to two large branches of life, and which has apparently been saved from fatal competition by having inhabited a protected station.[7]

Given the absence of fossils from such animals at the time, his interpretation of the place held by the enigmatic platypus (a bizarre egg-laying mammal) or the lungfish (a South American swamp-fish with rudimentary lungs that is able to hide underground during lengthy droughts) is remarkably insightful. Darwin concludes with a hint of uncharacteristic lyricism:

As buds give rise by growth to fresh buds, and these, if vigorous, branch out and overtop on all sides many a feebler branch, so by generation I believe it has been with the Great Tree of Life, which fills with its dead and broken branches the crust of the earth, and covers the surface with its ever-branching and beautiful ramifications.[8]

Almost a century and a half later, these lines betray no hint of age. The image of life as a great tree is just as precise and useful now as it was then. Its many spreading branches illustrate the broad spectrum of biodiversity, and the fact that there is only one tree denotes the unity in all living things. Since *On the Origin of Species* first made its appearance, Darwin's key underlying idea, natural selection, has been bolstered and refined by much research in genetics, molecular biology and paleontology, and still it holds good. It is the most potent theory there is to explain life, or what John Maynard-Smith calls the principal unifying idea in biology, for it allows a satisfactory description of nature and is indispensible to the understanding of life in all its complexity. Moreover, on a more subjective note, Darwin's tribute to these "ever-branching and beautiful ramifications" seems to me equally justified.

In Montreal's botanical gardens, my search for the perfect specimen to represent Darwin's description of the tree of life is just a tad facetious. After all, it's nothing more than an image to help us understand things, a teaching tool. There is no problem with the general principle — each tiny branch stands for a present-day species, so the closer one gets to the trunk, the further back one travels in time to rejoin ancient ancestors. The same holds true for the ecological dimension, although it may be a bit subtler to comprehend: two small branches can be close to one another (like any two species, say wolf and caribou), but their last point of contact, that is their last common ancestor, might be far back on the trunk. Also, there can be only one route to reach a particular branch, or only one way to arrive at a given species. Such is the distinction between classifying biology and shelving books in a library — the latter allows for a variety of different approaches.

Every comparison has limits, of course, and it is important to spot them. The main one is the shape of the tree. Normally we like fine, upstanding trees, but the tree of life cannot be seen as a straight shaft whose peak is topped by mankind. This misleading concept leads to a false notion prevalent in antiquity and the Middle Ages, that of the *Great Chain of Being*. In fact, evolution is not a journey to be marked out in graduated steps from point "A" to point "X," then eventually to humankind. It has no end. Nothing was foreseeable from the outset, not even humans. In the tree of life, humans can only be considered one twig among many, and indeed might never have come to exist at all. This imaginary tree is by no means a straight evergreen or conifer with humans perched on top.

Might it then be some majestic tree with balanced foliage, say an oak or an elm, a beech or a maple? Not really, since the great classifications of life aren't like the evenly distributed branches of these trees. In fact, the various kingdoms of life have been diverging over a period of 3.5 billion years. Some thicker branches (phyla) have yielded a large number of medium-sized branches (classes, orders), and others very few; some of these have rapidly subdivided into many smaller branches (families, genuses) and twigs (species), and others nothing at all. Through all this, many great losses have occurred and entire branches have simply fallen by the way and disappeared, while others, such as mammals, have grown quickly, despite their relatively late emergence.

If you think about it, the tree of life probably best resembles a bush more than a tree: a full, thick, wide one — not especially pleasing to the eye. That day in the botanical garden, I began to eliminate the most attractive trees, and finally, empty-handed, abandoned my quixotic quest for the tree of life altogether. Back at home, as I put my bike away in the garage, I saw it. There before me was a tangled honeysuckle, its foot hidden by last autumn's dead twigs and leaves — maybe my search had yielded a tree of life after all!

Before Darwin, naturalists had already established a system for cataloguing living things. From the 17th century, they had developed reasoned approaches based on the comparison of living samples with collected fossils, and this morphological analysis is still a key method practised by taxonomists the world over. It marks the beginning of the scientific period. By analyzing the forms and structures of organisms, scholars of the time established lineages, which they presented within the explanatory frameworks they then possessed. Before Darwin, they relied on the notion of *species fixity*, as in the creationism promoted by religions: in other words, the transformation of species according to the "life force" theory of Jean-Baptiste Lamarck. Darwinism then provided a general theoretical framework that was convincing, one that corresponded to preceding observations and could be tested by them.

One striking example was found in the skeletons of different vertebrates. Place a lizard's claw, a dog's paw, the arm of a human and the fin of a whale side by side, and you will see that, whatever the shape or function, the actual skeletons are virtually the same: the bones are similar and *placed in the same order*. One can then suppose that these animals had a common ancestor which bequeathed them a distinct skeletal structure. This idea of a common lineage, together with natural selection, allowed Darwin to perceive unity in a number of phenomena that his predecessors had observed, both in nature and in deep geological layers.

Still, it was quite some time before the Darwinian theory of natural selection obtained pride of place, for reasons ideological as well as scientific. It was revised and amended by further post-Darwinian studies. Put simply, there were contributions from genetics, notably population genetics —

referred to in the 1930s and '40s as *the new synthesis*. Then in the 1960s and '70s another theory, known as *genetic neutralism*, and finally in the '80s, the *Theory of Punctuated Equilibria*. Nevertheless, the number of vigorous debates has simply served to reinforce Darwin's arguments, and it is safe to say that nowadays no serious biologist doubts that natural selection is the driving force of evolution, though divergences remain as to the speed and manner with which it occurs.

Nevertheless, 20th-century genetics has certainly shuffled the deck. When the nature of DNA was discovered in 1953 and the key to the way genes worked shortly afterwards, the outlook changed. It was utterly ground-shaking to discover suddenly that the entire spectrum of life — plants, animals, bacteria — resembled one another at the molecular level. The following material, although demanding for the lay reader, will give some of idea of this outlook and lay the groundwork for the rest of the book, but some new concepts must first be introduced.

We have already mentioned that the genetic code is universal, but what does it actually do? First, it's a link between the languages of the four-part foundation of DNA (A, C, T and G) and of proteins — the workhorses of any living organism — (twenty amino acids). Of course, the combination of two languages and an intervening translation code may seem cumbersome, yet it works well. Every protein in every known living being is constructed according to this code — *only* this code, every protein being a distinct amalgam of the twenty basic amino acids, hence the root unity of all living things.

Put another way, DNA is the enduring thread that connects us to all our ancestors, even the most remote. This thread, invisible to the naked eye, in fact ties us to every living organism: from the most primitive bacteria to humankind, via the ancestor of fish, a primitive amphibian, a mammal-reptile, a first primate, and an *Australopithecus*. There is absolutely no break at all in this genetic transmission line, its "genetic capital" being carried by the double helix that is the universal information-storage system. Obviously, DNA and the genes it transports undergo mutation and selection within each species and even each individual member, but this thread of life is unique and irreplaceable. It has appeared only once in the history of life on Earth.

The importance of all this to the way we see living things nowadays is of course enormous. To gauge this, let me refer to the second famous "D" in the world of evolution: Professor Richard Dawkins of Oxford University. In his 1986 book *The Blind Watchmaker*, he points out that before the age of molecular biology, the only way zoologists had to determine the lineage of animals was by the large number of anatomical similarities among them. Now, however, the treasure chest is open wide, and once-vague impressions of parental links could almost be considered statistical certainties. Each protein became a "sentence," a chain with amino acids as "words" that could be "read" in various organisms. We must assume that two very similar proteins then had to come from species that were very close cousins. Dissimilar proteins came from distant cousins. The exact difference between two animals can actually be measured by the number of different words in a sentence.

Parallel to this is the belief that each protein proceeds at a relatively constant speed, regardless of the animal group (there are, of course, exceptions, but these can be set aside for the moment). Thus the differences between similar proteins in distinct animals provide a good indication of the time elapsed since their common ancestor existed. This is the underlying principle behind what is called the *molecular clock*. This time machine allows us to determine not only which animals are closest cousins, but also when their joint forebear lived.

Dawkins describes the method's power thus:

> Not all molecular sentences in all animals have yet, of course, been deciphered, but already one can walk into the library and look up the exact word-for-word, letter-for-letter, phraseology of, say, the haemoglobin sentences of a dog, a kangaroo, a spiny anteater, a chicken, a viper, a newt, a carp and a human. Not all animals have haemoglobin, but there are other proteins, for instance histones, of which a version exists in every animal and

> plant, and again many of them can already be looked up in a library. These are not vague measurements of the kind which, like leg length or skull width, might vary with the age and health of the specimen, or even with the eyesight of the measurer. They are precisely worded alternative versions of the same sentence in the same language, which can be placed side by side and compared with each other as minutely and exactly as a fastidious Greek scholar might compare two parchments of the same Gospel.[9]

He concludes that DNA sequences are in fact the gospel of life, now decoded for all to see.

In the concrete reality of research, all this can be rather complicated. There are many devices and techniques for comparing species directly via genetics. A simple example would be to compare a protein, human cytochrome c, with its equivalents in the rhesus monkey and the horse. There is only one difference between the gene sequences (the 66th of the 104 amino acids in the protein) in humans and monkeys, but 12 between humans and horses, and 11 between horses and monkeys.

If one assumes a roughly even rate of change from one branch to another, one may hypothesize dates for the divergences among these three animals. It can be represented by at least three different trees, but which is the right one? This is chosen by a principle known as *parsimony*, which aims at finding the lowest number of independent changes in the cytochrome c protein over the evolutionary course.

Of course, it would be easy to say we might have guessed humans would come out looking closer to apes than horses, because anatomy and paleontology predispose us to it. Research works in the same way: the number of possible phylogenic trees becomes enormous, and from the outset, we must eliminate the least probable ones to concentrate on the others. This still requires computers with specialized programs. Methods of genealogical reconstruction based on the distances between sequences (taken two at a time) are available, but there are others. These include using the number of mutations (substitutions of DNA bases, insertions, deletions) that affect each site (relative position) in the sequence.

Before leaving these rather demanding theoretical points, we need to answer an obvious objection: "O.K., that's fine, but can we rely on just one protein to trace the lineage in all living things?" The answer is no. Let's quote Dawkins once more:

> Molecular information is so rich that we can do our taxonomy separately, over and over again, for different proteins. We can then use our conclusions, drawn from the study of one molecule, as a check on our conclusions based on the study of another molecule. If we are worried that the story told by one protein molecule is really confounded by convergence, we can immediately check it by looking at some other protein molecule.[10]

The English biologist is referring to the phenomenon of *convergent evolution* — different organisms finding the same solutions to the same problem by independent routes. In that event, though statistically rare, comparison of several proteins allows us to remove the ambiguity by eliminating the coincidences. Finally, among instruments, we must mention the increasing number of genome data banks: as of 2009, the genomes of about 150 bacteria, a score of plants and about 50 animals (mice, fruit flies, earthworms, zebra fish and chimpanzees) had been decoded. The number has since doubled: the panda, the cow, the bee and the sea urchin are among those that were added by the end of the year.

At this point, it makes sense that the sport we refer to as *molecular phylogenics* would involve finding or defining the real linkage between organisms with a battery of sophisticated methods. Darwin's image of a single tree still works, but in reality, researchers have been building a great many of them to indicate the evolving lineages according to the scales of families and genera rather than larger groupings, if only because of the immense number of details involved. In fact, we're faced with a universe resembling Russian dolls, with each model fitting inside another. The dynamic effect is like the zoom lens of a camera.

So let's begin with the twig (lost somewhere in the middle of our thick, wide and tangled bush) that is humans, which is attached to the single branch (among many others) that represents mammals. First zoom-out: we can see our most remote vertebrate ancestors, the earliest tetrapods. Next come our invertebrate cousins and the ancestors we share with them. With the second zoom-out: new animal branches appear, notably anthropods (spiders, shellfish and insects). A little further, and we can see coral, jellyfish, sponges and the tiniest marine organisms.

By now, that tiny last twig that is humans has become just an indistinguishable dot. We discover the expansion of microscopic fungi about a billion years ago, then another zoom-out and we see the first multi-cell organisms, and microbes — bacteria, viruses and archeobacteria — as they propel their distant offspring in long filaments toward the top of the bush into such density that, by comparison, the animal kingdom looks like a miserable little chicory tree.

Nowadays we can see pictures of this fabulous story in books that synthesize the subject in rich illustrations. Yet the computer is better equipped to show the extent and dynamics of natural history, and the arrival of the Internet has made all of it accessible to the uninitiated.

Surfing the Web can give us a perfect idea of the present state of science. There are some very specialized sites full of scientific jargon of course, but these can be hard to digest or lead us down some sort of digital black hole. Still, one Internet site is "worth the visit," as the *Michelin Guide* might say: the international Tree of Life project at http://tolweb.org/tree. Their tree of life, a robust, spreading bush that looks a bit like a dishevelled aquatic plant, is remarkably successful. One can zoom in and out at will, as mentioned above. The scientific information is of the highest quality with commentary from researchers of serious repute. From section to section, the site allows us to discover some amazing parental links within given families or genera, and measure the evolutionary distances as well.

Returning to the surprises of modern phylogenics, did you know, for example, that we are closer to flowering plants than we are to the *E. coli* bacteria we carry around in our intestines, or that we are closer still to mushrooms? This may seem frivolous, but there can be significant consequences when it comes to fighting fungal infections.

There are many more examples that could be cited, but this gives us an idea of modern phylogenic methods and the importance they can have. We have come a long way from the simple matter of classifying things, or ancient quarrels about where such-and-such an organism fits into the animal or vegetable kingdoms. It is clear that the changes brought about by the methodologies relying on DNA still have further to go and relate to the origins and the diversity of life, as well as the status of the larger groupings of living things. Lastly, they also relate to the way this primate we call humans see their cousins and even their own role on the planet.

In the following chapters, we will use two especially interesting case studies to illustrate this new way of looking at life. One concerns the origins of that important mammal group, whales. We'll see how the tricks of phylogenics, combined with classical paleontology and the new science of evo-devo, have brought about a major reclassification. First, though, let's take a closer look at the old question of classifying living things and show how molecular approaches have led us to renew our inventories of biodiversity.

And What's Your Bar Code?

IT'S EARLY ON A MAY morning in 1992, and we are walking a path through the Monteverde Cloud Forest in Costa Rica: four Canadians, fresh arrivals, making their way through the rainforest, disoriented and sweating profusely. Still, despite appearances, the plan is simple and not really that adventurous: to film a report on medicinal plants. Our guide is Manuel, a young local botanist who points out trees and plants for us to examine. So we do. I put down the tripod we've been lugging — and it seems to weigh a ton — then we film, I take notes, and we leave.

At one turn in the path, we meet two men coming toward us: the all-too-obvious foreigners loaded down with backpacks. Tourists? Well, not exactly. They're American botanists, Thomas Daniel and Franceso Almeda, on an expedition for the California Academy of Sciences. They are putting together a plant atlas of Costa Rica. Pure, unhoped-for luck! We set up an interview, right there in the middle of the trail. Daniel explains what they have been doing:

> When we come upon a plant specimen that we don't know *a priori*, the first thing we do is document the location; then we identify the specimen by its scientific name, not the popular local one, but its scientific Latin term: name and nickname. This means we first have to examine its morphology using *dichotomic keys*: considering it as having — or not having — a particular structural characteristic. Then we go on to the next, and the next and so on.

In fact, he admits a specialist can easily identify family or genus on sight, "but a more in-depth study is often needed to identify species and subspecies, especially in an environment as rich as this."

These comments take us to the very foundation of a scientist's work. The actual process of species classification is still similar to that invented by Carl Linnaeus (or Carl von Linné) in 1750, a system built on a hierarchy of seven nested descriptive categories. Humans, for instance, would be described this way:

Kingdom — Animal
Branch or Phylum — Chordates
Class — Mammals
Genus — Homo
Species — *Homo sapiens*

By convention, scientists regularly use Latin for genus and species. Thus *Homo sapiens* must be human, just as *Catharantus roseus* has to be the Madagascar periwinkle, or *Acipenser fulvescens* the lake sturgeon.

The fact that this branch of science is still in use does not prevent the taxonomy from evolving or names used in the past from being replaced by new ones, the reason being that classification is not its only purpose — and it has left numerous researchers, including Darwin himself, somewhat indifferent. In fact, he contented himself with saying Linnaeus' system was "very practical." Its other purpose is to establish phylogeny, the evolutionary history of species, the position of each branch on the thick bush that is the tree of life. This is a far more complicated task, and one in need of modern techniques. In turn, the use of such techniques leads us to classifications that reflect evolution as thoroughly as possible, little by little, as our knowledge increases. Taxonomy is in fact a very meticulous discipline often embroiled in passionate discussions by its practitioners, but hardly understandable to the uninitiated.

Nevertheless, the importance of classifying is obvious to all. Naming organisms in order to recognize them and comprehend their inter-relations in time and space, naming them to describe the richness of life and inventory biodiversity, as well as to conduct basic and applied research, is clearly indispensable. These are, of course, activities with enormous economic impact: how are we to manage fish stocks or animal hunts, for instance, if we cannot even tell them apart? One might go on to mention species that harm crops, or dangerous species that must be spotted and controlled before they do harm.

In this light, the main difficulty is not in the final stages of identification. Most plant and animal collections nowadays have been standardized and computerized for easy and universal access. The difficulty lies in the time required for the preliminary stages: specimens must be collected in the field, and sample fragments must be prepared for examination. Next, an expert specialist may have to be consulted, especially if there is room for doubt, and all of this is lengthy and costly. It should be possible in the initial stages to identify animal and vegetable species; it is here that we come upon molecular biologist Paul Hebert (formerly Hébert).

Hebert, a fiftyish, round-faced man with alert eyes, considers himself a "rather impatient" man. An evolutionary specialist at Guelph University, he became extremely irritated at having to waste time identifying animal and vegetable species. He and his colleagues had spent hours doing it with the microscope and the naked eye. Then, one day, as he tells it, he had a flash of insight. If all food products can be identified by a bar code, there must be something similar one can do for living things, a specific genetic code as it were. It took six years for him to prove his idea valid and convince funding bodies to support his project, the Barcode of Life Initiative.

Now, after launching the consortium at Guelph in May 2005 to identify all the fish on the planet, he can at last be assured of the project's continued existence. The core of his network at Guelph combines — via Internet connection — several international consortia for the identification of mammals, birds, fish and insects, as well as a nationwide Canadian network. Every day, hundreds of specimens reach the lab and promptly receive distinctive "bar codes" analogous to the rapid identity codes we find on foodstuffs. The goal is to build a centralized reference bank which aspires to become *the* biodiversity library the world over. The unabashed aim is to shift the inventory of life forms into high gear. It is an urgent enterprise, for we currently know only about one-fifth of these, and many are vulnerable and in need of protection before it is too late.

So how does one find the equivalent to a bar code for animals or plants? First, a distinct sequence-segment of DNA for each species has to be isolated. For Paul Hebert, the obvious choice was to use mitochondrial genes, since these are simple to obtain. These tiny bodies are the energy providers in each cell. Better yet, they are plentiful, easy to isolate, and have specific genomes of their own. These genomes are capable of rapid evolution, so that two divergent species will also have distinct sequences.

Hebert first worked on two mitochondrial genes, cytochrome c oxidase I (COI) and cytochrome b. He demonstrated that a small section of DNA near the beginning of the COI gene was all that was needed to be sequenced in order to identify any animal species with certainty: the 648 primary bases of the gene, in fact. This then is the piece of DNA he regularly sequences first in any given sample, after which he enters it into the central reference bank on the Internet. "The bottom line is really quite simple," he concludes. "Despite centuries of naturalistic observation, we still do not have a tool for identifying or reading living things. We don't have a method that we can transpose generally to all living things, and this is precisely what the DNA bar code will give us."

Simply put, the Barcode of Life Project translates a long-known reality that we had not managed to pin down. Scientists already knew that each species possesses its own unique signature in the four-letter language of DNA: A, C, G and T. Now this is accessible to us by browsing the reference bank, which is publicly available on the Internet. If, for example, we want to find the genetic ID file for a domestic cat, a few clicks of the mouse (non-edible) and the entire bar-code sequence shows up on our screen, starting with TACTCTTACTTTTATTCG. Then, the African elephant (*Loxodonto africana*) starts with the initial twenty letters: AACACTGTATC TATTATTG.

"It can't possibly be that easy," you think? Want something a little harder? Let's try identifying an animal based on its COI sequence using the project's standard method. This is the sequence you will see on your screen:

> GGTATATGATCAGGTCTAGTTGGAACCGCACTCAGTTT
> ACTTATTCGTGCAGAATTAGGTCAGCCTGGGGCCCTT
> TTAGGAGATGATCAGTTATATAATGTGATTGTAACTGC
> TCATGCATTTGTTATAATTTTTTTCTTAGTTATGCC
> TATAATGATTGGAGGGTTTGGAAACTGATTAGTTCCTT
> TAATACTAGGAGCTCCTGATATAGCTTTCCCACGCCTAA
> ATAACATGAGTTTCTGGCTTTTACCTCCTGCCTTAC
> TTCTTCTTTTGTCCTCAGCCGCTGTTGAGAGAGGAG
> TTGGGACGGGATGAACAGTCTACCCTCCTTTAGCGG
> GGAATCTCGCGCATGCCGGGGGTTCTGTTGATCTCGCT
> ATTTTTTCGCTGCATCTTGCTGGTGTCTCGTCTATTT
> TAGGGGCTGTAAACTTCATTACAACTATTATTAATATAC
> GATGGCGAGGAATAGAGTTTGAGCGACTTCCACTA
> TTCGTTTGGTCAGTAAAGATTACTGCAATTCTTCTTC
> TTCTTTCTTTGCCTGTTTTRGCAGGAGCTATTACAATG
> TTATTAACAGACCGAAACTTTAACACTGCATTCTTTGAC
> CCGGCAGG

Another two clicks, and here is our animal's identity with absolutely no possible doubt. The data bank registers 100 percent similarity, so we can be sure it is a mollusc, a gastropod of the Cypraeidae family, *Luria tessellata*, a very pretty seashell found mainly in Hawaii.

The Barcode of Life Project does have its critics, though, who point out that this method won't work for some types of anthropods, such as spiders, as well as numerous insects and the entire plant world. For the latter, COI evolves too slowly, and Paul Hebert is fully aware of this, responding that in questionable instances, another gene or combination of valid genes can be found. Solutions are at hand, he maintains, and it is just a matter of applying oneself to find them. His confidence is equally unshakeable when one worries that the genetic data assembled is largely confusing and therefore hardly useable. Promoters of the project point out that in 2004, using the Barcode, they were able to identify flawlessly every bird in North America in a mere six months. That in itself is remarkable enough, but in the process, they also discovered some new species. "After we analyzed 130 species of North American birds," he explains, "we identified four that were unknown to us. By extrapolation, that means approximately 500 new species across the world!" The study concludes that so-called "twin species" — that is species that morphologically are very similar to one another — can nevertheless be distinguished by their bar codes.

Buoyed by this success, the biologist believes his project will win over the sceptics in the end. "There will always be a place for traditional taxonomists," he points out. "The goal is to allow these specialists to avoid having to identify each and every sample individually and concentrate more on evolving new knowledge in their domain." The argument appears to be well-founded. For instance, there are very few specialists working full-time on identifying specimens of certain families of flowers, fish or lepidopteras sent to them from around the world. Furthermore, the specimens are frequently from developing countries, which often have the greatest biodiversity but don't have such highly qualified personnel. Seen thus, the genetic bar code method is especially interesting, because it is both rapid and standardized. In laboratories equipped with sequencing apparatus now widely available, every sample gathered in the field can be analyzed. Even the smallest are sufficient: an insect limb, the tip of a feather, a section of fin, or a few mammal hairs.

In the October 2008 issue of *Scientific American*, Paul Hebert and his colleague Mark Stoeckle, of New York's Rockefeller University, conclude their interim report thus:

> We view barcoding as creating a map of DNA diversity that will serve as a framework for subsequent detailed study. Just as the speed and economy of aerial photography caused it to supplant ground surveys as the first line of land analysis, DNA barcoding can be a rapid, relatively inexpensive first step in species discovery. The "ground truthing" will take more time. But

> linking these approaches will produce an integrated view of the history and present-day existence of life on earth and help to shepherd life's full magnificence into the coming centuries.[11]

Concretely, the Barcode of Life Project can also lead to a new generation of useful diagnostic instruments for those who work with plants and animals in the field. The idea is to adapt medical bio-chip technology: "At first, we can imagine a portable DNA extractor connected to a portable computer to allow us access to data banks," explains Jeremy deWaard, director of the Guelph lab. Next, remote communication technology can be used to give an instant reading of the sequence in any sample taken directly from nature.

Any layman can one day make use of this. So, does that wretched mosquito that just bit you carry West Nile virus or not? Even with a good entomology text close by, you might never know. Now imagine having a little electronic gizmo that you insert into a mosquito's leg that will tell the whole story in an instant! It's the same thing for butterflies, ferns and mosses all around: in fact, any living thing you can think of. This wealth of knowledge isn't required for one to admire the life around us and sense its magnificent oneness, but it does tell us something about where humans have come to in their search to understand it all: another body of learning that gives this lucky, big-brained animal an enormous responsibility as the protector of it all.

Hippos in the St. Lawrence

ALL SUMMER ALONG the 200-kilometre North Shore of Quebec, you can be part of a non-stop festival. Every morning on the piers of Baie-Sainte-Catherine and Tadoussac, a smallish crowd of tourists piles onto different types and sizes of boat to go whale hunting. It's a peaceful expedition, of course, where the only shooting is with cameras and the only capturing with binoculars. The soundtrack records cries of delight whenever a pod of belugas appears, sometimes directly alongside the boat, maybe one of the whales even giving it the once-over from the corner of its eye, or when the elongated head of a common rorqual whale suddenly emerges, followed by its dorsal fin and its supplely weaving twenty-metre body. A rarer spectacle is when a blue whale, the largest animal ever created (larger than a brontosaurus, heavier than a small family of elephants) serenely breaks the surface, its breath propelling a nine-metre column of water upward before it dives again with a huge slap of its giant tail fin.

The real miracle is that this armada of boats — everything from Zodiacs to 200-passenger cruise ships — has not driven the whales away entirely. There are rules, of course, for boating and observation, yet the volume of tourism just seems to keep growing. The recent creation of a national marine park at the mouth of the Saguenay River has likewise helped protect these sea-going mammals, but still the balance is a delicate one. Recent studies, moreover, are now indicating that the disturbance brought about by 300,000 visitors a year is more serious than thought at first. The animals tend not to feed for as long as normal when bothered in this way. In addition, the local whale population, once heavily hunted, now has trouble reproducing when the St. Lawrence is polluted by industries upriver. The mouth of the Saguenay, whose chilly fresh water meets the salt water of the St. Lawrence, is the critical habitat for a number of species. Rorquals in particular find a concentration of krill here greater than any in the Atlantic Ocean, and belugas, the smaller white whales locally known as "marsouins," also like to frequent these fish-rich waters.

It's been a long time since I whale-watched, and the swarms of people in recent years worry me, although I understand the excursions from Tadoussac are still enjoyable. In June of 1982, I was lucky enough to see the belugas from a Zodiac in the company of two young biology students. At the time, a lively debate was underway concerning the total number of belugas in the St. Lawrence, a debate which succeeded in splitting the research community in two. In the estuary and gulf, there is a small beluga population descended from more northerly populations, but now genetically separate. While the belugas are not considered endangered worldwide (particularly because of their abundance around the Arctic Circle), the same does not hold true for the southern group. It has been decimated by the hunt, which was legal until 1972, and by the ravages of industrial pollution. Some remarkable studies by the biologist Pierre Béland and his collaborators have shown in more recent years that toxic organochloride residue has accumulated in the organs of these small whales and caused various ailments, including cancer.

This, however, was not known in 1982 when I went out in the Zodiac. In fact, we did not even know the exact size of the population: was it merely a few hundred in poor health and not reproducing well, or could there be thousands of them in good health and multiplying rapidly? This had to be settled. At the time, the inventory was taken from the air with a methodology that left much to be desired. My two guides might have been only second-year biology students, but they firmly believed in direct observation, so binoculars in hand, they camped out on the riverbank and noted the daily movements of the belugas around Pointe-Noire. Some days, they would head out in a borrowed Zodiac to photograph them. Because of a slow leak, they could only stay out at sea for an hour or two, but still felt quite safe.

When I arrived, we climbed straight into the boat. It was five in the afternoon, and there were hardly any waves. A few minutes later, as I looked back to see how far we were from shore, I heard them shout, then shut off the motor. Suddenly, less than 300 metres away, white backs appeared and disappeared again. It was fascinating: twenty or so females, completely white, with their blue-grey young ones and dark brown newborns. We couldn't see their blowholes, but the instant we turned off the motor we could hear them distinctly, and we could see their large heads with the "melons," large fat-filled protuberances that allow them to transmit and receive signals underwater. As the Zodiac drifted along, the belugas made no move to distance themselves, but seemed to form a circle with some of the younger ones chasing one another inside it, and the females perfectly calm with their newborns clinging to their sides.

Drawing of beluga with young by Calvin L. Hooper, 1884

"Click, click, click," my guides rattled off photos like machine-gunners. The belugas were still blowing off noisily, and I remember catching a whiff of fish in the air — *well, what did you think they ate with those lovely sets of teeth?* Oddly, the other detail that comes back to me is something I wasn't expecting at all — the flexibility of their necks. When their sleek, five-metre bodies stopped in mid-dash, and they turned their heads back toward us, I was struck to see their necks bend, and the huge heads that seemed to be watching us, slightly inclined and obviously mobile. With one of them,

I thought I detected a slightly curious and vaguely ironic glance from the corner of its eye. Quickly, the circle dissolved, and they leisurely swam back out to sea. We followed them for a while, but the spell was broken, and the Zodiac was deflating visibly, so it was time to head back.

In the Middle Ages people still believed that whales were huge fish, although some ancient scholars were more clear-sighted. Aristotle, for instance, concluded that these air-breathing animals that nursed their young could not possibly be fish, and he classified them separately. The question was further resolved in the 18th century, and naturalists henceforth treated them as marine mammals within the order of cetaceans and including the sub-order of teeth-bearing (odontoceti) and baleen (mysticeti) whales.

More obscure was the mystery of their origins. Some claimed whales as the marine ancestors of all earthbound mammals. Others, not so: whales, it was claimed, were descended from earth animals and took the reverse trajectory. This is the hypothesis Darwin suggested in *On the Origin of Species* in 1859, although he offered no conclusive evidence for it. At the time, paleontologists were moving from one discovery to another, unearthing fossils of strange aquatic creatures, some having several points in common with cetaceans. The best known of these was the *Basilosaurus*, of which a partial skeleton was found in 1832 in Louisiana. The geological context suggested a very elongated animal with small feet, going back about 40 million years. A giant sea reptile some called it, others a primitive whale, basing their conclusion on similar fossils found in Europe. The profusion of fossils clouded the debate over the origins of whales. It was not until William Flower came on the scene that things became clearer.

A specialist in mammals who had made his name along with Thomas Huxley in his battle against Richard Owen, Darwin's sworn enemy, Flower had been the first to dissect a panda. He had examined many other animals in detail, but he was best known for his study of cetaceans observed on every possible continent. At the end of his life, he began to think seriously about their origins and in 1883 published a long article entitled "On Whales Past and Present, and Their Probable Origin." In it, he left aside the discussion of *Basilosaurus* and fossils of other extinct species. He broke with contemporary discussions and concentrated on the comparative anatomy of whales and other extant mammals.

In his survey, he begins with a discussion of general shapes and proportions, then remarks that whales have retained the bones and ears of earth mammals, though not using them in the same fashion. Next he moves on to the vent nostril, anatomically similar to nostrils, although he maintains that whales cannot smell. He then demonstrates that each whale fin contains the same thirty bones as the forequarters of earth-bound mammals, with the same articulations, although all but one are fused in the whale. He also demonstrates that whale skeletons retain vestiges of what was once a pelvis and a femur (thigh bone) found on earth ancestors and quadrupeds. He goes on in perfect Darwinian fashion to look for a plausible lineal connection taking earth animals into the sea. He proceeds by elimination — excluding, for example, seals as the closest cousins to whales because of the way they use their hindquarters.

From these observations, he draws a plausible ancestor, leaning toward a semi-aquatic animal with a thick, smooth skin:

> We may conclude by picturing to ourselves some primitive, generalised, marsh-hunting animals with scanty covering of hair like modern hippopotamuses, but with broad, swimming tails and short limbs, omnivorous in their mode of feeding, probably combining water-plants with mussels, worms, and fresh-water crustaceans, gradually more and more adapted to fill the void place ready for them on the aquatic side of the borderland on which they dwelt, and so by degrees being modified into dolphin-like creatures, inhabiting lakes and rivers, and ultimately finding their way into the ocean.

After which, he maintains, the way is clear for the diversification we can observe today:

> Favoured by various conditions of temperature and climate, wealth of food supply, almost complete immunity from deadly enemies, and illimitable expanses in which to roam, they have undergone the various modifications to which the cetacean type has now arrived, and gradually attained that colossal magnitude which we have seen was not always an attribute of the animals of this group.[12]

The idea that whales are "sea hippos," ancient land mammals having invaded the marine environment, was revolutionary at the time and garnered Flower modest success for a while, but was quickly forgotten at the beginning of the 20th century. The main problem was the same one that haunts paleontological studies: what has become of the intervening forms? What was the common ancestor, and when did the two groups diverge? How does one go from archeocetes (as Flower named the groups of primitive whales such as *Basilosaurus*) to modern whales? Why do only some whales have teeth or baleens (long, vertical blades made of keratin which allow the filtering of food)? Curiouser and curiouser!

Now let's skip a few steps. There have been some breakthroughs by paleontologists — and we shall return to them — but it was above all molecular biologists who kicked off the debate again in earnest in the late 1980s, when they compared whale genes with those of other mammals. Using the modern methods referred to in the second chapter, they constructed phylogenic trees and arrived at some convincing conclusions. In their analyses, whales came out as related exclusively to artiodactyls (ungulated mammals with an even number of hooves, like pigs, cattle, camels, deer and even hippopotami). In fact, a number of studies have placed them in that category.

In 1997, John Gatesy of the University of Arizona published more detailed findings. These showed that, of all artiodactyls, whales are closest to hippos. Well, Flower might just be skipping and dancing in his grave. At last, 114 years after he had claimed just that, it appeared whales were descended from animals resembling hippopotami who had gradually ventured into the oceans.

So, thus ended the great debate! Well, not quite yet, because until paleontologists actually dug up the real ancestors of this mammal group and followed the trail of their expansion over the planet, the story of whale evolution could not be completed. Until the late 1970s, these fossils were still missing. Soon though, a young American researcher by the name of Philip Gingerich was to sort out the puzzle pieces.

Having spent a few years digging up ungulate and rodent fossils in Wyoming, Gingerich found a new hunting ground: Pakistan. Although it was a long way from his home base at the University of Michigan, it was still a logical enough place to search for origins. When the Indian subcontinent collided with Asia, thus forming the Himalayas, the Sea of Tethys had receded. Gingerich found himself exploring the northeast banks of this huge extinct sea, and in 1979, near a village called Chorlakki, his team unearthed a series of arteodactyl and rodent fossils from rocks about 50 million years old. In the middle of it all was a coyote-size skull with a fine set of teeth. Beneath the skull was a bony structure resembling the protective shell of a whale's inner ear, and held together by a small S-shaped bone that only whales possess.

Gingerich, with Donald Russell of the Musée d'histoire naturelle de Paris, called this discovery *Pakicetus*, and they considered it a primitive whale living in deep waters 49 million years ago. In the years that followed, the American paleontologist returned and discovered more *Pakicetus* fossils, but soon Pakistan fell victim to ethnic disturbances and was closed to researchers. Gingerich then headed for the Western Nile in Egypt, where a large number of *Basilosaurus* fossils had been found, as well as those of a possibly related animal named *Dorudon atrox*. However, it was impossible to date these fossils exactly or to make useful anatomical comparisons.

Then one day, Gingerich found a more complete *Dorudon* skeleton; near the 48th vertebra was a small pelvis, and next to it a femur, a knee and a tibia — a complete paw. Such a member was too small to serve as a leg, and so it must have been the vestige of ancient hindquarters. For Darwin, such remainders were some of the best proofs of evolution; if God had created

Reconstruction of Pakicetus *extrapolated from skeletal remains.*
Illustration by Carl Buell, http://www.neoucom.edu/Depts/Anat/Pakicetid.html

all animals in their present forms, he argued, why would He make such frivolous mistakes?

So, we were dealing with a mammal whose forebears had lived on *terra firma* and then "returned" to the water. The evolutionary outline was becoming more detailed: from *Pakicetus* (a long-headed mammal swimming with four paws) came *Ambulocetus* (with a longer body and shorter paws), then *Rodhocetus* (discovered in 2001 by Gingerich's team in Pakistan, and resembling a large dolphin with a small astragalus bone — characteristic of arteodactyls). Finally, we reach the *Basilosaurus* and *Dorudon*, precursors of modern whales. Now our evolutionary bush begins to emerge, and everything is in place for the whale's true "lineage" (a somewhat misleading word we can nevertheless hold onto). What a remarkable expansion it was, from 55 to 38 million years ago, and from the western to the eastern shores of the Tethys Sea, as well as the North American continent.

HIPPOS, PIGS AND WHALES

So when did hippos split from whales in all of this? One needs to refer to the work of Jean-Renaud Boisserie and his team, as cited in the periodical *PNAS* for 2005. After carefully examining the paleontological and genetic data, they arrive at the conclusion that hippopotami and whales had a common water-dwelling ancestor 55 million years ago from which two groups sprang: the first cetaceans (leaving land and becoming totally aquatic) and a group of four-legged animals (anthracotheres) which disappeared less than 2.5 million years ago, leaving behind just one descendant, the hippo. This detailed and convincing study thus eliminates all close lineage between whales and the pig family (suidae) and at last reconciles fossil and molecular data.

It is September 2005, and I am on Quebec's North Shore of the St. Lawrence to tape a report on global warming and coastal erosion. At the stroke of noon, here I am, on a dune facing the Gulf of St. Lawrence, abandoned by my TV colleagues who are off filming the coast from a helicopter. I wait by the car with only a sandwich for consolation. It's a beautiful day, and the sea is calm. I gaze at it, because there's always a sense of freedom in the limitless horizon, but in fact, there's nothing to see: no boats, no surfboards, no swimmers. This is it, the end of the world. A century ago, there was practically no one here but aboriginal people, and the closest village, Mingan, is fifteen kilometres to the east. Before me is a wasteland as deserted as the one behind me, a forest of scrubby spruce trees with a sandy trail through it that is hard to find, despite the helicopter pilot's directions.

Well, nothing happens. Then all of a sudden, I see a dark line moving out on the ocean, and it's a whale! Dorsal fin like a sickle: there is no mistaking it. The first thing that comes to mind is an ordinary rorqual, but given the size of it, I begin to wonder. It may be far off, but it's still only seven or eight metres long. But then I decide it must be a small rorqual, swimming eastward in a straight line across the surface. I look for its spout and blowhole, but it's too far off for that. (Later, Richard and Prescott's guide will convince me that it is a small rorqual: their spouts are often hard to make out, unlike the larger ones.)

All at once, his back arches, and I can clearly see his dorsal fin in the centre. Then he dives again without ever showing his tail. Search as I might, he does not seem to resurface anywhere. Well, that's the end of that short encounter with a fellow mammal. At this latitude and this time of year, it is not an unusual thing in the Gulf of St. Lawrence. Beside the abundant belugas and small rorquals, we can sometimes see humpback, blue, black, bowhead, black pilot and, more rarely, sperm and orca whales.

This lone and unobtrusive individual near Mingan had baleens, not teeth, although it wasn't possible to get a close look at his impressive row of them. According to Richard, there are "270 to 348 yellowy-white baleens 15 to 30 centimetres in length." They are a characteristic of those cetaceans known as mysticeti, whereas odontoceti, such as belugas, have teeth. Research has shown that the two groups separated about 15 million years ago, plenty of time for them to adapt themselves to different prey and expand into all the world's waters, insofar as those were accessible to such large creatures. Thanks to the "marine highways," they are the only sizeable mammals, apart from humans, to appear in all of the world's larger regions.

It is interesting, by the way, to note how traits passed down from the common ancestor of both hippos and whales have helped them with later adaptations. Without living specimens, we will never know for certain, but suppose these primitive whales inherited thick, smooth skins, a heavy layer of subcutaneous fat, and highly developed hearing: all characteristics of the modern hippopotamus. Obviously, these provide a distinct advantage in extreme climates like the Arctic. This may be pure speculation, but let me venture an additional observation: after reading an article that estimated the milk of a mother beluga is eight times richer than cow's milk, I went on to find information on the milk of *Hippopotamus amphibius*. I discovered that it too was very rich, certainly much fattier and more filled with vitamins than that of animals slightly further removed but nevertheless sharing the same environment: boars and African warthogs. Coincidence? Perhaps not.

After all this, why is it so important to establish cetaceans as the closest relatives of hippos? For specialists, there is no question at all. They clearly and fully grasp the importance of understanding the history of mammal evolution and of the race for knowledge. In this quest, many of them display an existential joy, a rewarding intellectual jubilation. If we hold dear the advancement of science, the answer to the question is simple. From patient and rigorous research comes a fresh vision of the evolution of whales, one which overthrows established genealogies and offers new perspectives in the understanding of life forms. No mean feat. This alone can raise delightful professional discussions, as well as drawing in a new generation of young researchers.

There is another level of response to the question, which is simple enough and helps broaden the discussion. Outside the circle of biologists and paleontologists, it is clear that the diversity of life forms — and the threats they face — can only be understood in an evolutionary framework.

The job that Darwin began 150 years ago is not complete and still commands great attention: first, because of the increasing pressure of human action on the world's ecosystems; second, because of the challenges to evolutionary science by fundamentalist religious institutions — currently a widespread phenomenon in the United States. When creationist thinking (based on the idea that God created the world with a determinate number of non-evolving species) is on the upswing in the most powerful country in the world, there is cause for concern. This is particularly true of its avatar *Intelligent Design*, which continues to make gains among the younger generation, promoting biblical narrative to the detriment of evolutionary science in schools.

To put this in perspective, let's go back to a CBS poll in October 2005, in which 51 percent of Americans believe that God created humans in their present form; 30 percent thought humans had evolved, but only under God's supervision; a mere 15 percent believe that Cro-Magnons evolved without divine intervention. With the American heartland in full anti-evolutionist battle mode, the consequences may be lasting. Among Canadians, 58 percent accept evolution, while 22 percent think that God created humans in their present form within the last 10,000 years, and 20 percent are unsure, according to a poll by Angus Reid Strategies, conducted in July 2008.

To the question posed earlier, I would like to respond on a third level, more existential perhaps, but in my opinion, likely to affect each of us. Essentially, the kinship among animals inevitably causes us to reflect on our own situation as well, and on our common fate with the animal world. Personally, I realized this a very long time ago when I found myself face-to-face with a dolphin.

It was near a Breton village at the tip of Raz, where my family originated. Two friends and I had set out in a motorized canoe in search of plaice, fish we caught with an underwater gun or a long trident. Our basic equipment was a mask, snorkel and fins for the depth of three to four metres, and we alternated between surface and bottom fishing. All of a sudden, a dolphin burst out of the depths — looking huge to me. It froze for a moment, then circled around us. I froze, too, but I wasn't afraid for long, because we'd already spotted him a short way off from our canoe. Although they usually

AN UNFAIR BARGAIN

Since 1986 and an international moratorium, large whales cannot be hunted anywhere in the world. Yet Japan, Norway and Iceland continue to do so. Japan alone kills more than 1200 of them annually under the guise of "scientific research." Moreover, for fifteen years, Japan has been promising aid to small countries in an attempt to buy their votes and reopen the hunt. In June 2006, it succeeded in mustering a majority in the International Whaling Commission to vote for a resolution condemning the moratorium. Defenders of the *status quo ante* have several options open to them, but this is still a definite step backward for conservation and shows how fragile the protection of this species really is.

stay far out to sea, they sometimes come into shallow bays. This one was obviously curious and watched us for a while, then circled us once more, and with a powerful tail stroke set out to sea.

It was then I realized how well he was adapted to his watery surroundings, and I was not. He breathed with lungs the way I did, and had to surface from time to time for more air. Still, he could stay underwater much longer than I was able, and his streamlined body allowed him to slide through tens of kilometres of sea water without getting tired, and at a speed I could never hope for. I became cold quickly, even in summer, which is why I wore a drysuit, but not him. Underwater I was practically deaf, but he could hear sounds my ear could never detect. I'm sure he could see much better than I could, too, with my ridiculous mask. He could navigate by sonar and communicate with his peers through the clicking sounds he emitted. Of course, he could also dive to depths I could never manage, even with much more sophisticated equipment.

Still, somehow we were obviously connected. The shapes of our skeletons, the functioning of our organs, our reproductive systems were quite similar. Two species of large mammals reigning over a marine habitat populated by hundreds of smaller ones, and why not slip each other a wink in passing? I'm sure he enjoyed the same plaice we were after, mackerel even more, along with the shrimp that local fishermen harvest every day.

In this dance of life, he and I found ourselves at the apex of the food chain. *A priori*, we had few fears of predators, but even so, we cannot call ourselves equals. His family and species are no threat to mine, yet all cetaceans are threatened by humans. Dolphins may no longer be hunted along the coasts of western Europe, but they sometimes die in fishing nets. In Japan, Peru and Chile, among other places, they are still hunted, despite their protected status. As a matter of fact, our impact on the hippo is even greater. They too are killed for their meat, and their habitat is constantly shrinking through deforestation. In Madagascar, at least three species of hippopotamus have disappeared in recent history.

Ultimately, understanding how we relate to other life forms speaks to our sense of responsibility as humans. Our history of predation — even deliberate massacre — of species and their habitats is dramatic, but not unavoidable. Becoming aware of the connections and interdependence among all living things seems a perfect way to begin acting responsibly: first individually, then collectively. The fact is simple: by virtue of our common heritage, our position in the living world and the environment we share, our fate is intimately and permanently tied to cetaceans and hippos. Knowing this, one must see them differently. Here's to you, cousins!

PART TWO

All Parts Included: Some Assembly Required

The Fly and the Butterfly Tell All

ON A HOT JULY afternoon in Vaucluse, we find shade under a large bush in the garden. It's the season for making apricot jam. Yesterday, I fetched two crates of apricots from a nearby farmer. The fruit is ripe for a nice thick jam. With a device formerly used as a vice for making frames, we will mix the apricots with crushed almonds. We use an old recipe that involves alternating layers of apricot halves and sugar and letting them soak together.

We've patiently prepared this, having put the container in a cool spot, except that I've forgotten to throw out the fruit we previously rejected as damaged or overripe, so here I am with a small pile of fly-and-bee-infested rotting fruit. Some flies are much smaller and lighter-coloured than others, and their larvae must have already been in the crate that I didn't throw away after hauling it out of my neighbour's basement. There aren't too many of these, but they're definitely fruit flies.

For an instant, I blink and imagine a long line of mutant flies, some with red eyes, some with extra wings, and others with feet where the antennae would be. Nope, the ones I actually see all seem normal, though I can't spot every detail. Better be careful. Maybe a little too much bubbly last night? No, actually, but I know where these hallucinations are coming from: they're a throwback to Bateson and Goldschmidt's mutant flies, the "promising monsters."

In fact, the genetic basis of animal formation was discovered about a century ago, while studying mutations in the humble fruit fly (Drosophila). The important name here is William Bateson. Fascinated by the mutations he discovered in these flies, the Englishman continued studying them persistently. It was he who invented the word "homeotic" from the Greek "homos," meaning "similar." He applied it to mutant organisms in which one part of the body has been transformed into another: say, a leg instead of an antenna on the head. Bateson wanted to show that these oddities of nature were at the root of evolutionary change.

Half a century later, the German geneticist Richard Goldschmidt took the idea a little further, postulating that these monstrosities were the beginnings of a new species: "promising monsters" he called them. We now know that these exceptional cases are in fact poorly adapted anomalies that natural selection will eradicate through their failure to transmit genes. Thus, the idea of such things as "promising monsters" has no basis, although curiously it has paved the way for modern embryology and the role of development in evolution. Via these mutants, we have discovered the genes that control the shape of animal bodies. Seventy-five years later, the subtle mechanisms connecting apparently dissimilar animals came to light. In this new view of life, as we shall see, we owe a great deal to the lowly fruit fly, so tiny and unlike us in appearance, yet so close. Here's to evening jam-making sessions, and to the Drosophila that has taught us so much!

"Promising monsters" remained a scientific mystery for a long time. The animal world turned out to be full of them, and they were awful: one-eyed sheep, wingless chickens, six-legged frogs. Yet we did not know what caused the aberrant forms. Soon, however, scientists were able to create them in the lab by exposing embryos to chemical substances or by physical manipulation. They thus understood that certain genes had a specific role to play, but it was not until new lab techniques appeared in the 1970s that things started to become clear. Researchers were able to identify two series of genes grouped on the third chromosome of the fruit fly, which seemed to control its development from embryo to adult.

The first series, the *antennapedia* complex, contains five genes affecting development in the front part of the body, while the second, the *bithorax* complex, contains three affecting the rear portion. In 1983, names were given to these sequences of 180 nucleotides (*homeotic box*), and the corresponding *homeotic genes*. The proteins produced by these genes regularly contain 60 amino acids.

To sum up: a series of genes on a single chromosome governs the development of the body of the notorious fly, aligned with the axis from the antennae to the tip of the abdomen. The corresponding proteins being very similar, one might suppose that they have a common function. But what is it? It was then realized that they had the same activating mechanism as certain proteins already familiar to us. They acted as *genetic switches* installed on the DNA: on, they activated the expression of other genes; off, they did not.

And there you have it. The function of proteins in the homeotic box is to act as switches as the animal is developing. This is why they manage the formation of structures such as antennae, eyes, legs and thorax, despite their occurrence on different genes. Contrary to what everyone believed, Walter Gehring's team at Basle started to find homeotic boxes in all sorts of animals: insects, amphibians, vertebrates as small as a mouse or as large as a human or a cow. All of them, however, needed to have bilateral symmetry. The sequences of 60 amino acids were so similar (scarcely two divergences between the drosophila and the mouse, for instance) it was astonishing. This meant that organisms that had branched off from one another over 200 million, even 500 million years earlier — in the case of insects and large mammals — had nevertheless kept the *same homeotic genes*. In fact, even the order in which genes were activated in the course of development was the same, regardless of how distant the species were from one another, as was the way in which these were used. These regulatory genes, as they are known, revealed the true depth of commonality among all living things.

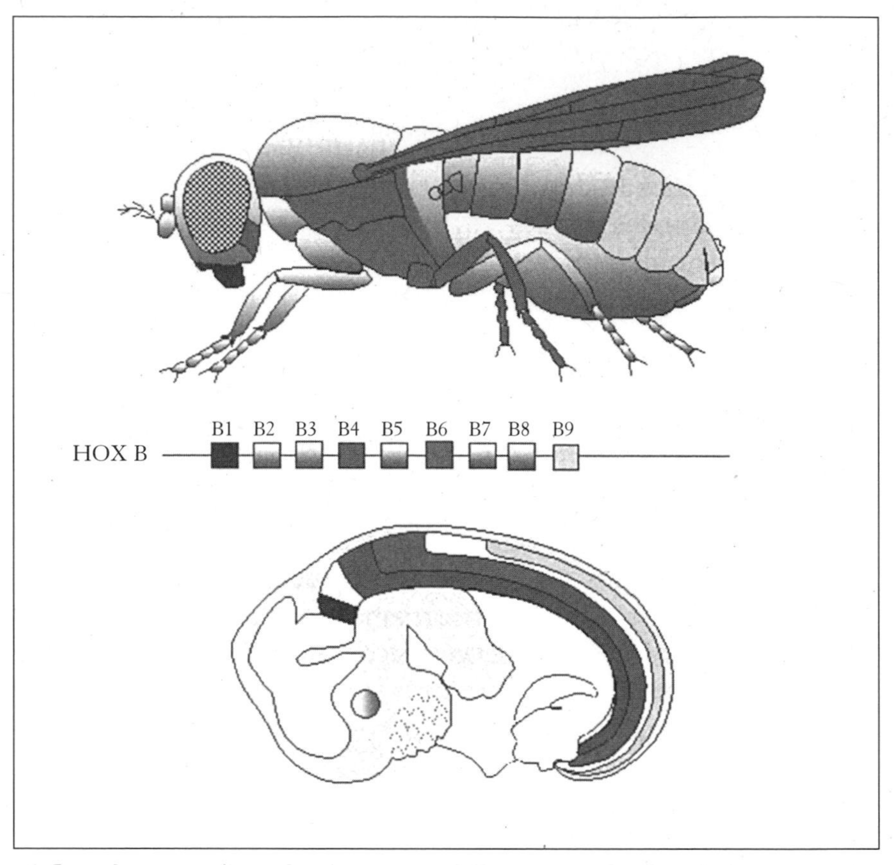

A fly and a mouse (in embryo): two very different animals whose body structure is affected by the same homeotic genes. Illustration by Alain Gallien, Académie de Dijon, http://svt.ac-dijon.fr/schemassvt/sommaire.php3

Nor are these genes the only ones that fit this category. In the wake of these discoveries, Gehring's researchers uncovered a remarkable gene, the mutations of which caused flies to lose their eyes (the *eyeless* gene). It was intriguing, for it resembled a gene known to cause a diminution in the iris when it is mutated (*aniridia*), similar to a gene known as *small eye* in mice. So was this another structural gene? It seemed hard to believe, since there are about forty types of eyes in living creatures — too many differences between insect and human eyes, for example, for us to believe they have the same basic structure.

Yet in a few elegant experiments, Gehring and his American competitors shattered this dogma to bits. Not only were *eyeless*, *aniridia* and *small eye* genes similar and widespread in the animal world under the name PAX6, but they were indispensible in the formation of all these different types. The slightest mutation, and the eye is either badly formed or not formed at all. There was no further possible thought of forty independent inventions, but one and one only, likely happening in a common ancestor to all these animals. It would be equivalent to hearing forty different jazz interpretations of the same piece with the same chord charts, yet forty different scores!

Soon, in fact, PAX6 was joined by a whole series of "little brothers," genes needed for the proper formation of other organs, such as the heart — the fly has one, even though it possesses only a tubular heart: *tinman* it is called (a nice bit of imagination for once), from the character in *The Wizard of Oz*, who has none. This gene would be the precursor of the genetic system needed to form a complex heart. One last example: the *distal-less* gene, which, when mutated, leads to modification of the extreme members. Fruit flies no longer have feet, fish fins atrophy, urchins have shrivelled appendices, as do rabbits, mice, and in fact, every lab animal that was tested.

The revelation that encapsulates all this work is of extreme importance: overall development, especially the formation of organs and members on the body's different axes is subjected to the action of a small number of highly powerful "regulator" or architect genes. Further, there is a basic genetic recipe for the creation of every type of eye, heart or limb.

The day after putting up our jam, we're off on a picnic nearby. Our home is in a corner of the country on the slopes of Mont Ventoux. When it's unbearably hot as it is today, one of our favourite cure-alls is to climb the mountain and breathe the cooler forest air. This is something we do regularly, and we love this walk from the spring at Grozeau through a little pine grove on the mountain at Piaud. The ground's very rocky, and you need good shoes for this, as well as a hat and lots of water, but it's well worth the effort.

The trail is steep at first and winds through the pines at Alep and then through green oak. We climb at a slow-donkey speed, ears humming from the insect song, careful where we put our feet. Jean-Henri Fabre, the region's great naturalist, and someone I read a bit at a time whenever I'm here, sees this in the trails:

> . . . an unending layer of sedimentary rock that breaks off in flakes that rattle away with an almost metallic sound under our passing feet. The cascades of Ventoux are a constant babbling of rocky bits rustling like its own special brook.[13]

Soon we're at the first clear ridge, a huge rock sprinkled with spindly bushes and junipers. This is really the time for a break. Nearby, white butterflies flit among the flowering plants. Marie-Ève unties the scarf from her hair and tries to catch one. At first, no luck. They're too fast for her. Then, with one last throw of the scarf, she bends down and finds a large white butterfly lying on the ground under it. I go over to it carefully, and although I know practically nothing about them, I'm able to identify it: white wings with bright red spots circled in black. It has to be an Apollo. I'm sure of this, because once again, Jean-Henri Fabre has described them so well. Then I begin to have some reservations, since this one is at a height of only 1,200 metres, well below the peak where he locates them. A few days later, however, a local naturalist reassures me, saying *Parnassius apollo* also occurs at a lower altitude.

Time for a closer look at this one. It is magnificent in its white livery with lines and black marks on its upper wings. On its lower wings, there are large carmine marks which catch the eye, two small upper ones on the lower edge that look like eyes. This one has an incomplete right wing, which seems to have been partly torn off, only the black circle of the upper mark remaining.

"Why is it missing part of its wing?" wonders Marie-Ève. "I mean, I didn't do that to it. Do you think it can still fly?"

Here, I'm on familiar terrain, and I can play the know-it-all and give my ego a boost. "A bird or a lizard must have done that," I tell her. "See those eye-shaped marks on its wings? Those are bait to fool predators by drawing attention to non-essential parts of its body, so if something pounces and tears off a bit of its wing, it'll survive and still be able to fly, but an attack on another part of its body would be fatal."

Spots of Apollo butterfly (Parnassius apollo)

This is something one can learn from any biologist, but Marie-Ève, being young and curious, chimes in: "O.K., but how did it make those markings?" Terrific question, although this time the answer isn't so obvious, because it has to do with both "how" and "why," and for that one needs to understand much more. I was a little quick with my answer before to Marie-Ève, so now I'll have to do better. You see, if one knows enough about the "why" and the "how" of butterfly markings, that's because of a new specialization where genetics meets evolution and embryology: evolution-development, or evo-devo as its friends call it.

At this point, we need to duck back into the lab and talk about the discoveries of biologist Sean B. Carroll of the University of Wisconsin-Madison. He discusses these in detail in Chapter 8 of his book *Endless Forms Most Beautiful*. In the late 1980s, Carroll and his colleagues identified several genes responsible for the formation of wings in fruit flies. Since insect wings have appeared only once in evolution, it seemed reasonable that identifying the corresponding genes in the butterfly might show if some of them contributed to creating the characteristic markings. Carroll was doubtless not the first to wonder about this, or to use the "all we have to do is . . ."

shortcut, but in this case, Carroll was lucky, because it did lead to an answer to the enigma of butterfly markings.

First, a side note: the beauty and diversity of these creatures come from two innovations that occurred after they separated from other insects in the course of evolution. These are scales (the name lepidopterans comes from the Greek for scale, *lepis*, and wing, *pteron*) and the coloured, geometric shapes on the wings. It is thought that originally scales were modified hairs found on numerous insects, each one stemming from a single cell. The shapes on the wings come from the progressive development of cells in sections juxtaposed in parallel from the base of the wing to the outside: in each section, there is a subdivision edged with veins, and similar distributions of cells are repeated, so that, seen from above, an alignment of bands or markings occurs. Finally, each scale is one colour only; it is spatial repetition of scales of differing colours that gives the overall impression.

Now, back to Sean Carroll's research. The first step was a relatively easy one: yes, the butterfly did have a similar series of genes for wing-creation to that of the fly. Next, the researchers tried to find out when and how these genes were active in the insect's development. Since the wing is formed out of a flat disc of cells that become enlarged throughout the different stages of development, it occurred to them to track these cells closely, which can be done by relatively sophisticated means in the lab. Result? The surfaces, as well as the edges of wings, are determined by the same genes in both the fruit fly and the butterfly. This means the construction mechanism ("architectural plan") is one and the same.

That's not all. The scientists also discovered the mechanisms by which these genes express themselves and are unique to butterflies. They were able to establish that each band or marking could change shape, colour or size independently of anything else. The genetic orders appeared to be individualized, and it only remained to identify them. Let's hear this from Carroll:

> I will never forget when my technician Julie Gates called me to the microscope to see the most stunning pattern of beautiful spots in the caterpillar wing discs. We saw two pairs of spots on each disc precisely where the eyespots would appear a week later in development, in the positions that Fred Nijhout (an entomologist colleague) had defined as the foci of eyespots. Fantastic![14]

The most amazing surprise was the fact that markings depended on a single gene that the team had been studying for two months — in fact, one they already knew very well. We too were acquainted with it a few pages earlier — *distal-less*, the one needed for the complete formation of limbs.

In the buckeye butterfly they studied, the *distal-less* retained its initial function, expressed as it was in the extremities, but in this particular butterfly, and in all those that bear markings (as Carroll's team later demonstrated), this gene had a further role — precisely that of creating the markings. But why was that?

> The gene acquired a new switch that responded to the specific longitude and latitude coordinates of these spots. These *distal-less* spots always form exactly between the two veins and along the outer edge of the wing. The precise and reproducible coordinates of these spots tell us there are tool-kit proteins active at these positions that flip on the switch in the *distal-less* gene.[15]

In other words, like a good pet, the regulating gene had learned a new butterfly trick. A new genetic switch allowed it to create a shape on the wing that it could not form in other winged insects.

All that remained was to spell out exactly how the ocelli are formed. More complex than simple markings, they resemble eyes with a coloured centre, then a circle of another colour and a dark outer border. Research indicated that, if *distal-less* was responsible for the formation of the centre of the false eye, completion of the entire form relied on two other genes, *engrailed* and *spalt*. Each of these marks the beginning of a different circle and determines the relative distribution of various coloured scales in the circle. And — guess what — it turns out these two genes also have an additional switch in the butterfly that allows them to play a further role. Three homeotic genes each acquire a new switch: that is all it takes to create the fake eyes that fool predators. How beautiful we humans find them even beyond all that!

Nor is this the sole example of gene regulators. Particularly well documented is the case of the Mexican axolotl salamander. There are dozens of species, some stuck at the larval stage and living an infantile life to the end. Others, by contrast, transform normally, the difference shown to occur from the mutation of a single gene regulator, which in this case produces a hormone. Furthermore, this gene is known to have been activated and deactivated several times in the history of the species. In other words, the infantile axolotl is either an earlier adult form or a more recent one, and there is no telling if they will change again at any time.

Once again, a basic theme returns: nature doesn't dream up every recipe from scratch. She often goes with modified versions of old ones and fiddles with the basic materials.

For famed French biologist François Jacob, natural selection operates more like a do-it-your-selfer than an engineer:

> A handyman doesn't know right away what he's going to make, but he collects whatever materials come his way, even the oddest things: bits of string and wood, old cardboard — all of it might eventually come in useful. In other words, he makes use of whatever he finds lying around and finds a use for it. An engineer, however, never sets to work until he has everything he needs at hand: tools and materials. The home handyman, though, will make do with leftover bits, and often what he makes does not fit any master plan.[16]

The beauty of evolution is its open, modular system. Successive inventions emerge from a small number of basic rules. The changes happen when, for instance, architect genes like *distal-less* happen to "bump into" a new function that integrates a new module, a functional segment of DNA from some other organism. At times, another is dropped instead, or the two are fused, and something new occurs. It is a series of small touches without any predetermined plan, just a shuffling of the possibilities. This is how Jacob summarizes the general principle which applies to butterfly markings:

> Evolution proceeds like a home handyman, who for millions and millions of years, slowly reworks his piece, with a retouch here and there — on and on — cutting here, lengthening there, never missing an opportunity to adjust, transform and create.[17]

Of Finches and Their Beaks

IT BEGINS UNREMARKABLY enough in Vancouver in 1960. His name is Peter, and he is British, a twenty-three-year-old doctoral student in zoology at the University of British Columbia. She is Rosemary, twenty-four, and is in charge of practical assignments in the Biology Department, also British and a passionate ornithologist. What happens next is quite simple, also very human: far from home, they fall in love one hot, humid autumn. Rosemary, here for just a semester, decides to wait a bit before returning to her own doctoral studies in Upsala, Sweden. Her stay in Vancouver is going to change everything, leading to one of the longest and most productive collaborations ever between two biologists.

The Grants made their mark with notable field studies of a very ordinary bird: the finch. Their work provides elegant and incontrovertible proof of the theory of evolution. In the words of the Batzam Prize Committee

(2005), "The work of the Grants has had a seminal influence in the fields of population biology, evolution and ecology."

Despite winning many other world honours in recent years, and giving much of their time to their Princeton biology students, the Grants have not changed much at all. They are up early every morning, headed straight for their modest offices, teacups in hand. Although they no longer spend three or four months in the Galapagos every year, they remain active in the field and have collaborated with two Harvard geneticists to identify the genetic foundations of beak development in finches. These are amazing discoveries, and the foundations are far simpler than expected, while still perfectly in accord with Darwin's ideas. Although one may sometimes quibble with the master on certain details — including finches — the Grants have provided still more genetic support for his ideas on the evolution of species.

The adventures in science of Peter and Rosemary Grant actually get underway a few years after their meeting in Vancouver. Holding doctorates, they now find themselves in Montreal, where Peter undertakes a post-doc in zoology, and Rosemary deepens her knowledge of genetics. They have two daughters, and while the daughters are growing up, Rosemary does some high school teaching. Peter makes his grand entry into McGill University in a new full-time position, and Rosemary soon joins him on some of his research projects. Their careers are now launched, and plans must be made. They decide to tackle two fundamental questions: how do species come to be formed, and why do some animal populations change more rapidly than others?

To look for answers, they decide to go to the Galapagos. This archipelago, 1,000 kilometres west of the coast of Ecuador, is a geological and geographical wonder. There are 19 islands and 42 islets, most of them fairly recent formations. They are the peaks of a submerged volcanic range, whose origins are linked to a hot spot beneath the Nazca Plate. They are strung out east-to-west like a necklace, the oldest having emerged a mere 9 million years ago, so they are quite new against the backdrop of what has been called deep time. This origin has allowed them contrasting shapes and environments, each with distinctive flora and fauna that underlie their reputation as a small paradise. There are animals found nowhere else, such as giant turtles, both aquatic and land-based iguanas, lava gulls and flightless cormorants. There are few mammals, and no amphibians or freshwater fish. Land-based vertebrates are essentially reptiles and birds.

Among them, finches are numerous and can be found throughout the islands, hence easily observed. None other than Darwin himself offered the first scientific description, however imperfect, on the *Beagle* voyage of 1835. Contrary to popular opinion, he didn't stay long, and he was drawn by a lot more than finches.

Still, aided by his faithful servant, he did set some nets to obtain specimens: forty in all. However, he did not label them correctly, failing to note systematically where they were caught, and this initially prevented him from grasping their importance. A few years later, though, urged on by ornithologist John Gould, Darwin became aware that each species occupied a different island. Thus, in the long run, he was not oblivious to their importance. Their great tendency to diversify proved the instability of species, contrary to teachings of the time. Already, as early as eight months after his visit, his oft-cited notebook indicates the following:

> When I see these Islands in sight of each other and possessed of but a scanty stock of animals, tenanted by these birds, but slightly differing in structure and filling the same place in Nature, I must suspect they are varieties. . . . If there is the slightest foundation for these remarks the zoology of the archipelagoes will be well worth examining: for such facts would undermine the stability of species.[18]

Since then, time has allowed us to identify thirteen distinct species in four families of brown, insect- and grain-eating, active and very noisy passerines. All of the species share very similar plumage and shapes, but — and this is crucial and a delight to researchers — there are three principal distinctions: body size and, above all, the shape and size of their beaks. Each has a distinct ecological niche. A quick family portrait shows *Geospiza* and *Platyspiza* (the "large beaks") eat grains and perch in trees, whereas *Camarhynchi* ("small beaks") are arboreal and eat insects; *Certidae* ("warbler

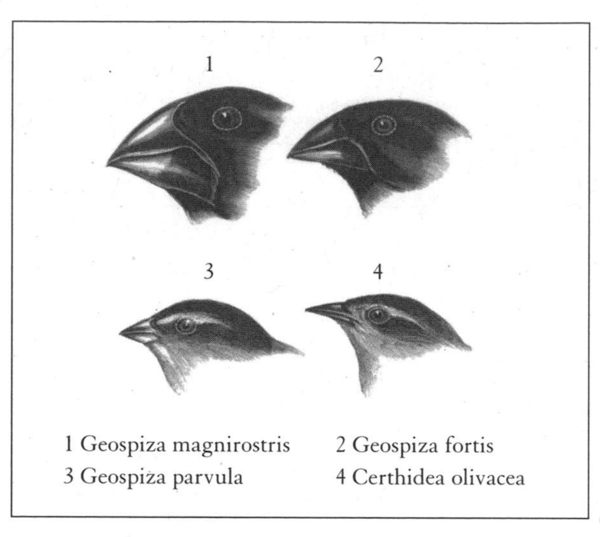

Drawing of four Galapagos finches' beaks by John Gould from The Voyage of the Beagle *(second edition, 1845) by Charles Darwin*

finches") and *Pinaroloxias* ("Cocos Island finches") are also arboreal and insect-eating, but have elongated beaks. Finally, the *Cactospiza* finch has a strong beak, which it uses in an unusual manner, tearing the bristles from cacti and using them to spear insects in the holes of tree trunks.

It is now 1972, and the Grants have found a natural lab to study the mechanics of evolution in birds. Three or four months a year for over thirty years, they camped on two of the islands, Genovesa and Daphne Major. The latter is the tiny peak of a volcano. Genovesa, on the other hand, is noticeably larger and greener. Uninhabited by humans, these two islands have almost no water fit for drinking, and only the rarest of researchers is even allowed to set foot there. Rules against contamination are strictly enforced (all food must be dried, fruit is forbidden for fear of non-indigenous insects, etc.).

One can imagine Rosemary and Peter living in small tents: frequently up before dawn to observe birds in search of food, putting out nets to catch them, taking blood samples, counting nests and eggs — later on tagging the young ones. During mating season, each female lays from three to five

FROM THE GALAPAGOS TO CANADA

We were long unaware of the Galapagos finches' origins. In 2001, however, the mystery was solved. Genetic research begun by the Grants concluded that the Grassquit genus *Tiaris*, and specifically the species *Tiaris obscura*, is the nearest living relative of Darwin's finches. An ancestor must have migrated from Central America 2.3 million years ago, proving once again that the vernacular name of "finches" is misleading. In any event, their "allies" as Darwin termed them, are numerous. Every region of Canada has several species, the best-known being the house finch of Mexican origin (*Carpodacus mexicanus*); the enlarged family includes familiar and beautiful birds like the grosbeak, blackbird and goldfinch.

eggs, and our biologists are allowed to sample one out of every three at the most, and only at the correct time, so as not to cause the nest to be abandoned. This permits them to document in detail the history of every single finch on the island: a spotless routine, a life for passionate nature lovers, an ascetic labour in a grandiose setting — all of this for thirty years!

From the outset, they noticed that the main selective pressure on finches comes from variations in annual rainfall. When plant growth is strong and bushes have plenty of leaves, there is a lot of seeds for birds. In dry years, there are far fewer seeds, particularly small ones. With environmental effects like the 1977 drought on Daphne Major, *Geospiza fortis* finches are forced to fall back on the less preferable large seeds, and thus the birds with smaller beaks tend to die out and reproduce in smaller numbers than those with large beaks.

A few years later, between 1983 and 1985, Daphne Major was colonized by *Geospiza magnirostris*, another large-grain eater. The large-beaked *Geospiza fortis* immediately suffered from this competition, while their smaller-beaked cousins made a major comeback. Thus the species *overall* has flourished through the generations, the result of its adaptability as indicated by the varying sizes of beaks. For evolutionary biologists, the surprise unveiled by the Grants was double. Until then, they had not known about the important variability of a single essential feature, beak size and shape, in wild populations. Even more, that such vertebrate groups could change so rapidly was astonishing: evolution right before your eyes!

What Darwin didn't know was that this adaptive capability resides in the finches' genes. During the 1990s, the Grants began to collaborate with a Harvard team led by Clifford Tabin and Arkhat Abzhanov — young aficionados hoping to succeed them. This new direction led them to identify the genes responsible for the shape and size of Galapagos finch beaks.

Asked what types of genes come to the fore and under what circumstances, readers of the previous chapter will want to exclaim, "Well, at the embryonic stage, of course!" and they'd be right. Here, once again, is the main proposition of evo-devo: the shape of animals evolves according to changes undergone in the development of the embryo. In this case, what would they be? Researchers have shown that the finches' larger beaks depend essentially on the expression of gene BMP4. The more it is activated at the embryonic stage, the larger the beak will be. This rule provides a simple key to explain a complex reality, for, as French Nobel-winner Jean Perrin commented, "the key to any scientific discovery is to explain the visible and complex by the invisible and simple." In fact, just by experimentally increasing the production of gene BMP4 in chicken embryos, the Grants and their colleagues obtained chicks with bigger beaks! In finches with long, slender beaks, however, another gene was found to be responsible, *calmodulin*. The sooner it is expressed in the embryo, the longer and finer the beak will be.

The coda to this symphony in beak major, if I may, is doubly stupefying. First, genes alone are sufficient to determine beak form in species, and not just a few dozen, as first thought. Second, robust as their genetic system

WHATEVER HAPPENED TO FINCH #5110?

On Daphne Major in 1981, the Grants for the first time observe an unusual member of the species *Geospiza fortis* (apparently from a nearby island), remarkable for its large size, unusually-shaped beak, and its song. They refer to it as 5110 and observe its "descent" over the years. After four generations, it so happens that its offspring die as the result of drought, with the sole exception of one small brother and sister. The story continues, however, for they mate and have offspring themselves, who in turn take family members as mates, until, while still maintaining all their distinct characteristics, they cease reproducing *except* among themselves. No doubt about it, the Grants are witnessing the reproductive isolation of one sub-population within another: the Holy Grail of evolutionary biology! In the long run, as mutations pile up in this population, it will no longer be possible for its members to reproduce with others at all, thus forming an entirely new species. In the face of these results, the Grants explain that sexual selection, and not drought alone, was responsible. Even with a vast choice of partners, females of this group are attracted to the large size and unusual song of the male descendants of 5110. In other words, "You're big, and you've got a nice set of pipes. Let's get acquainted."

may be, the evolution of species also depends on environmental conditions. What a fantastic, up-to-date and subtle lesson the Grants have served up!

Let us just footnote this by saying that this research, the true icon of the synthetic theory of evolution, is often under attack from creationists. First come partisans of the fixity of species, whose inspiration is drawn from the

Bible. Next come followers of the pseudo-science Intelligent Design (we'll come back to this in a later chapter), yet the arguments of both are basically the same. When they talk about speciation, they cling to the debate between *microevolution* and *macroevolution* in order to find the elusive little flaw. In so doing, they deform the real scientific debate. In order to answer both, it's worth talking about what distinguishes these two concepts.

Generally, microevolution considers modifications that have taken place within specific populations of a single species, such as the *Geospiza fortis* finch. As we have seen, these modifications can be brought about by natural or artificial selection, as well as by mutations and random "genetic drift." After generations, gene frequency evolves without yielding new species. Macroevolution, on the other hand, concerns changes leading to the formation of new species or divergence within large groups, say, mammals or reptiles. The time scale is different, of course, macroevolution being essentially a concept related to deep time, as in paleontology.

Can we call them opposites though? No, not at all. Modern evolutionists see no contradiction or even serious differences between the two. They are rather manifestations of the same evolutionary process. There is, to be sure, an internal debate on the *speed* of macroevolution; experts debate, sometimes quite sharply, the phases of stability and of rapid expansion that follow massive extinctions. The problem, however, is that anti-evolutionists charge into this *apparent* breach in order to criticize neo-Darwinism, claiming that it is proved by neither micro nor macroevolution. From this, they conclude (and the link is highly suspect), that evolutionism is in crisis or flatly false.

One example is an apparently well-informed article written for the Biblical Creation Society. The author, it should be mentioned, is a physicist by training and secretary of the society. After a correct summary of the Grants' work, he concludes:

> Natural selection is not a creative influence on organisms. It can only act on and mould what is already there. With regard to this point, neodarwinism links natural selection with sources of genetic variability (primarily mutations) to inject creativity into their vision of origins. The research with the Galapagos finches has cast no light on this aspect of evolutionary theory. No mutations are involved in the observed changes. There is no new genetic

information. The research has added to our knowledge of ecology and biological variation, but it does not provide any support for any theory of large-scale evolutionary change.[19]

He goes on to explain the creationist position: all finches are derived from a few species created by God, although he does admit to past microevolution in finches. Yet, in the same breath, he claims the Grants' research does not point to evolution in modern finches, and finally excludes it with the statement that nothing has been proved in terms of macroevolution, which he terms "large-scale evolution."

Objections of this kind are typically creationist and not trustworthy. The Grants have, in fact, shown clearly that physical aspects of the species (size of the birds, as well as beak features) change with the environment, the latter affecting survival and reproductive success, all in a shorter lapse of time than was previously thought. When, moreover, we witness the emergence of sub-populations that no longer interbreed due to geographical isolation from one another, we are, in fact, speaking of speciation, the formation of a new species. To maintain that there has been no evolution is false.

Furthermore, the Grants are not alone in having proved the appearance of new species "right under our noses." Over the past thirty years, the same phenomenon has been directly observed among bacteria, insects, plants, as well as mice, field mice, lizards, fish, and so on. That creationists try to sow the seeds of doubt is understandable, but here we are on very solid scientific ground; their position is made in bad faith and is patently absurd.

Stuck in the Mud and How to Get Out in under 10 Million Years

FIRST, YOU HAVE to go back in time, and for this your imagination is the best machine. We're now in the Devonian Period, extending from 410 to 360 million years ago, before continental drift gave Earth the look it has now. At this point, every continent is welded into one of two huge masses. The northern one is our concern for now, and we first need to locate Greenland, stuck between what will later be Europe and Quebec, only a few degrees north of the Equator.

Now imagine a swampy, tropical landscape crawling with all kinds of life, especially fish (with or without shells or jaws). On firmer ground, there are ferns, mosses, large plants — but no trees or forests as we know them — not even any dinosaurs, mammals or reptiles, just masses of arthropods: small, medium and large, and above all, giant scorpions. Welcome to the Devonian!

Here in the future Greenland, a strange creature is wading in the shallow estuary waters, something like a huge fish or salamander, but with fingers. Seen close up, there are other things that differentiate it from fish. It has gills but lungs as well. Some bones between the head and the pectoral girdle have disappeared, giving its small mobile neck more freedom of movement. Its somewhat flat head has eyes on top rather than on the sides. This is *Acanthostega* coming up for air.

Drawing of an Acanthostega *mock-up*
by Arthur Weasley

Now imagine a small local catastrophe, maybe a sudden and violent rise-and-fall in water levels. Our "fish" is dead, quickly buried in the mud or sand: the body decomposes but the hard bits are preserved, creating the *Acanthostega* fossil.

Back to the present. It's the summer of 1987, and a team of English paleontologists are on a fossil hunt in Greenland. Jennifer A. Clack of Cambridge University is in charge; she's a specialist in the first tetrapods (four-footed animals, or their descendants, with lungs — the frog, the penguin, the snake or the baboon, including us. There is no ecological cranny that doesn't have them).

She studies all the very first tetrapods, taking especial notice of the transition from fish to amphibians. It's an important transition because that is when vertebrates left the water to conquer *terra firma*. What she particularly wants to know is how it came about, when, and in which species.

The few weeks she has spent on the Celsius Berg formation in the eastern part of the island have been very fruitful, and the team has, in fact, unearthed some remarkable specimens of *Acanthostega*, our strange fish named by others before we had more than fossil fragments. At first glance, this animal is in no way unusual, but over the years, it gradually delivers one surprise after another. The series of articles Clack has published, along with her colleague Michael Coates of the University of Chicago, has become famous in the restricted world of paleontology.

Till then, we thought fish had walked out of water on sturdy fins and with lungs for breathing, then gone for a stroll on the beach. The best candidate seemed to be *Eusthenopteron*, a fish discovered in the 19th century

Eusthenopteron *fossil discovered in Miguasha, Quebec. Photograph courtesy of Miguasha Museum*

on the cliffs of a small Quebec village called Miguasha. This unpronounceable thing made the Gaspé village a scientific celebrity, and a charming local museum was opened. This creature could have been more than a metre long and swam in shallow water about 375 million years ago, according to the rocks in which it was found.

Paleontologists studying its skeleton were amazed by how much its members resembled those of all tetrapods, which is why they thought it was the first fish to have dragged itself up onto land. Many biology texts still show it doing just that as it searches for food or a swamp that hasn't dried up yet.

Surprise number one: our *Acanthostega* had a clearly aquatic lifestyle. It was a tetrapod, of course, but not the earth-walking kind. It had functioning gills and a vertically flat tail, with long fin rays on the upper and lower surfaces supporting a fin for effective sub-aquatic motion. Although it had lungs, the ribs were short and did not connect on the underside to form a solid thoracic cage. Trying to hoist itself ashore without such internal support would have caused the animal to collapse and thus crush its lungs. That's not all: the joints in its legs did not permit forward or backward walking, although they were ideal for paddling in water, but hardly effective for dragging it around on land. The elbow too was practically immobile.

So these first tetrapods were not earth creatures, but primarily aquatic. So much for the first dogma laid low by the *Acanthostega*, for until now, it was thought that fish had hauled themselves out of the water and later become earth-walking tetrapods. Thus legs and fingers appeared in response to the pressures of selection related to the aquatic milieu, but which ones? Only hypotheses are available to us. Analysis of the sediments surrounding *Acanthostega* fossils shows that it lived in the shallow and highly turbid water of estuaries, and evolved among aquatic plants which normally abound at this depth. The small hands probably helped it move about among them and grasp stems when it lay in wait to ambush its prey. Perhaps they also helped it "walk" at the bottom of the water, just as monkfish do with their fins modified into small, articulated legs.

Next surprise: fingers. The researchers patiently uncover the stone over the specimens: they see five fingers, then more. *Acanthostega*, it turns out, had eight on each foot! That splash you hear is another dogma going under. Hitherto, the leakproof "law of nature" had allowed tetrapods a maximum of five fingers or toes — pentadactyly. Some species had lost fingers or fused them together, like a horse's hoof or bird's wing, but very rarely had one found more than five. Now here was this little tetrapod proudly showing off all eight. Briefly, Clack and Coates wondered if they ought not to label this a congenital and individual deformation, just as humans are sometimes born with an extra finger.

At about the same time, however, other tetrapod fossils just as old were being discovered around the globe, and so validated the "normalcy" of the eight-toed foot. The bizarre *Tulerpeton* in Russia had six, and the *Ichthyostega*, likewise from Greenland, had seven. So it would appear that with the first tetrapods, nature played variations on the theme of fingers, and ours in the key of pentadactyly was just one, and the version that finally prevailed.

Now, it's all very well to note this, but it doesn't tell us how a simple fin can change into something with which we can play a piano. We know roughly where and when the first paws appeared, but still nothing about how. So let's get back to *Eusthenopteron*, the Miguasha fish, and especially its pectoral fin.

It was exceptional and had an arborescent skeleton, with a "branch" appearing at every new articulation, and all of them pointing to the same side of a central axis, toward the rear of the animal. The first evolutionists to study this question explained the transformation from fin to paw as a series of extensions with new small bones to form fingers. Thus each finger bone was associated with a wrist bone.

With *Acanthostega*, this didn't work. All of a sudden, here we were with three fingers left over and no additional wrist bone to complete the line to the central axis! The theory was beginning to look lame. Fortunately, Michael Coates was able to come up with another. He took up the daring idea put forward a year prior to the Greenland expedition by Neil Shubin and Father Alberch of Harvard University in one of those articles that so easily falls into oblivion. Fortunately, Michael Coates read it and remembered it when he first encountered the eight-toed paw.

Shubin and Alberch's theory went something like this: our ancestors' fins were constructed around a central axis from which all branches went in the same direction. Tetrapods retained this anatomical particularity: our radius and wrist bones still emerge from the same side, but the main axis does not extend directly into a finger. Instead, it takes an abrupt turn toward the extremity and forms a hook before what has become the wrist.

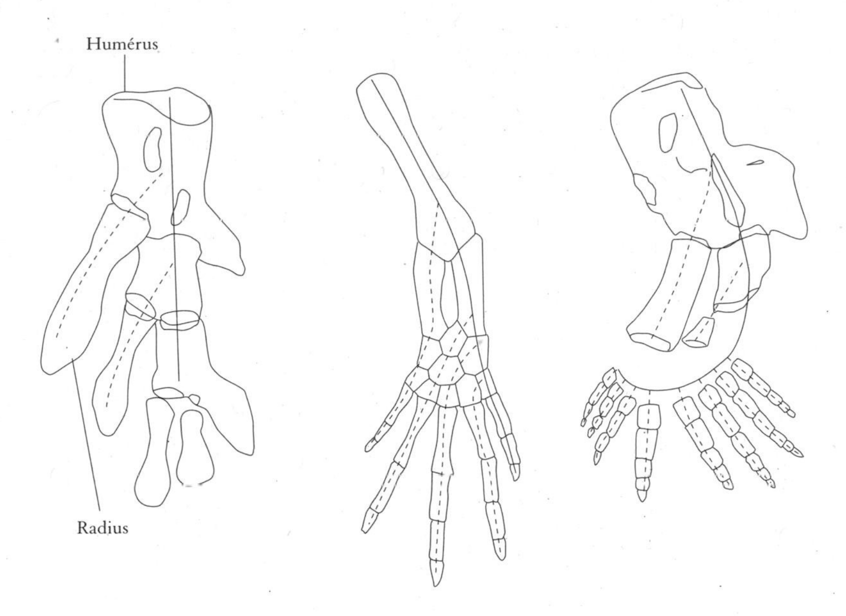

"Polydactyly in the earliest known tetrapod limbs" by A. Clack, Nature *347 (1990): 68.*

The branches were thus inverted to the other side of the axis, pointing toward the exterior and then forming fingers. How could that be? The authors believed that the number of fingers depends on the time the embryo has to allow branches to appear, one-by-one along the arc.

A bizarre theory? One might think so at first, but it's been proven in the lab. If the buds of future limbs in a frog embryo are exposed to a chemical product that slows the rate at which cells divide, the animal will develop not a complete miniature hand, but a normal-sized one without a thumb. With fewer cells in the buds, the frog is unable to grow the last digit on the arc. In nature, frogs have only four digits, because the arc stops growing before a fifth branch has a chance to emerge. On the other hand, the large Pyrenees dog and the St. Bernard sometimes have a fifth or even a sixth toe in front, whereas most dogs have only four.

On discovering the *Acanthostega* paw, Michael Coates began to make the connection with this theory. The first tetrapods had more time for development, he reasoned, which allowed the arc time enough to produce eight digits. Fortunately, this ancient tetrapod was discovered *after* Shubin and Alberch's article. Without their idea, we would have puzzled long and hard over this odd paw.

There is now a more recent chapter in the story, this time in genetics and embryology. Scientists studying the *Hox* genes (homeotic genes in vertebrates) also came up against aberrations when looking at the formation of limbs in tetrapod embryos. As we observed, *Hox* genes normally express themselves along the main head-to-tail axis in all bilaterally symmetrical animals. When *Hox* genes were first detected in action on these members, it was supposed that they would continue in the same way as along the vertebral column. So, they imagined, the head-to-tail axis simply became the shoulder-to-hand axis.

Not so. It was more complicated than that. In the bud preceding limb formation, they noticed instead that a *Hox* gene was expressed only on the side of the little finger, taking an abrupt turn and turning up again in the fingertips. This led to complete confusion amongst researchers — all except for Michael Coates. Having closely followed the work on developmental genes, which was to help him understand the evolution of limbs in tetrapods, he realized immediately that he had seen this structure somewhere else: it was precisely the curve of Shubin and Alberch's arc.

Coates showed this amazing concordance to the scientific community, maintaining that fossils, cells and genes all pointed to the same explanation. No one was familiar enough with the limbs to say exactly how the *Hox* genes contributed to the formation of this arc, but whatever the actual mechanism, the coincidence was too obvious to be ignored. In 1996, a team of American geneticists published a detailed portrait of limb development in tetrapods. Following the progression of twenty-three *Hox* genes day by day, they had discovered that they corresponded to Shubin and Alberch's arc more perfectly than anyone dared imagine. Fossil-to-gene, it all held up. In order to invert the branching out of the fin bones, evolution must have reversed the *Hox* genes — a clever piece of tinkering.

To sum up, have you wondered how our tetrapod ancestors made it out of the swamp? The solution is not so complicated after all. Starting with a rather special fish, *Eusthenopteron*, it takes prolonged dry spells and a little genetic craft. Time passes, a few hundred thousand years perhaps, but then when you are Nature, what does it matter? Lots of species die, and more step in to replace them. Within these species, the better-adapted individuals survive and pass on their genes. In the long-run, adaptations occur. Then, a few million years more, and among the remote descendants we find *Acanthostega*, which is finally equipped to climb out of the water. About 10 million years stuck in the mud and finally you're out! Now stay out.

Megouasag ("redcliffs") is the name the Mi'kmaq First Nations of the Gaspé Peninsula gave to this sector overlooking the Restigouche River where it empties into the Bay of Chaleur. Three centuries ago, the French colonists called it Miguasha. Walking along the beach on a fine late-September day, one can understand why it bears that name. The setting sun caresses the cliffs with its semi-warm rays, and for a few minutes, the eyes rejoice in the spectacle of the rock blazing in scarlet. The aboriginals must have been moved by it too.

Scientists found out about the site in 1842, when a geologist named Abraham Gesner — best remembered as the inventor of kerosene — came from nearby New Brunswick. His curiosity drew him across the river. "Fossilized vestiges of plants, fish and tortoises," read his report. The local farmers already knew these decorative stones, of course, like the natives before them. But no one knew the scientific value in these cliffs until Gesner discovered them.

O tempora o mores: that applies to scientists scrabbling in the rocks and earth too. Now, at sunset on the beach at Miguasha, there's a paleontologist gathering up the day's finds. Good finds for a good day's work. His five fingers gently stroke the small *Eusthenopteron* fin bones he'll be taking back to the museum. It's an exceptional discovery when you consider all the reasons this testimony to the past might never have been found: changes in the ecosystem hundreds of millions of years ago, sedimentation, fluctuations in the sea level, subsequent erosion.

One more stroke of the fin engraved in stone — of the two animal shapes spread before him, there's something of a genetic inversion, a random happening a long time ago in the history of life, and one with profound consequences.

There on the beach, the paleontologist puts the rock in his bag and heads slowly off to the museum. He thinks about these enigmas life has presented: their solutions here and now in the early 21st century; the scientific explanation that has spilled so much ink; and a better understanding of a crucial stage in the history of living things with their emergence from water and the great adventure of earthbound vertebrates. Through all of it, 375 million years of evolution is wrapped up in the day's reddening end.

The Thumb and the Baby Panda

IN THE LATE 1970s, as relations between the U.S. and China were beginning to thaw, the Washington Zoo received two giant pandas, which soon became the main attractions. I once went to see them and was fascinated by how these charming black-and-white "teddy bears" devoured their favourite bamboo plants. Perched squarely on their butts, they used their front paws to grasp the stalks one after another. Then they quickly stripped off the leaves and gobbled them up, carelessly throwing the leftover stalks anywhere.

What I hadn't noticed was actually most remarkable of all: pandas peel the stalks by slipping them between their thumb (apparently flexible) and the other fingers, and there are five of those. So, how come a sixth finger, and an opposable thumb as well, when this is only supposed to happen in primates? In 1980, the paleontologist Stephen Jay Gould at Harvard published a book entitled *The Panda's Thumb: More Reflections in Natural*

History, wherein he plumbed in great detail and depth the mystery of the panda's thumb. He even held it up as a beacon that stood out in the world of biology and among the great evolutionary strategies. When I went to the zoo, I knew absolutely nothing of any of this.

First, a detail: this thumb isn't as fully opposable as ours, being too short and rigid to allow that. It's not really a true thumb, as Gould explains:

> The panda's "thumb" is not, anatomically, a finger at all. It is constructed from a bone called the radial sesamoid, normally a small component of the wrist. In pandas, the radial sesamoid is greatly enlarged and elongated until it almost equals [the length of the first bones of the true fingers. It] underlies a pad on the panda's forepaw; the five digits form the framework of another pad. . . . A shallow furrow separates the two pads and serves as a channelway for bamboo stalks.[20]

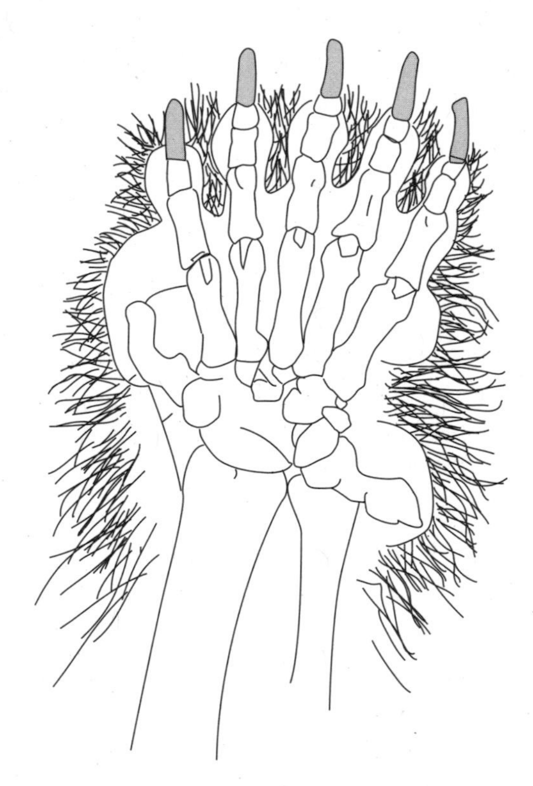

Drawing by D. L. Cramer from The Panda's Thumb
by Stephen Jay Gould

Thanks for the details, Dr. Gould, but where exactly does that leave us? The central theme is that these bizarre arrangements and odd resolutions are one more proof of evolution. He also touches on the idea of contingency or randomness in evolution, and includes the panda's thumb as striking proof of adaptation, or rather as François Jacob would say, "life's tinkering." Gould explains how, progressively evolving toward a herbivorous state centred on bamboo, "the panda is obliged to use available organs and choose this hypertrophied bone as a functional, but somewhat bric-a-brac response."

He goes on to mention orchids (which Darwin described so well), possessing large petals known as labella, used to trick bees by opening and causing them to fall into the nectar at their centre. Unable to get out again without rubbing up against the pollen, the bees on their next visit to another of these plants automatically become pollinators! One might be tempted then to consider the labella an elaborate system *fated* to ensure cross-pollination, but this was not their primary purpose; they simply evolved that way out of ancestral forms with entirely different primary functions. Their new purpose was handed to them accidentally.

Gould's book postulates that panda ancestors had a sixth pseudo-finger used for other things than peeling bamboo leaves. He supports its existence with paleontological evidence, for predecessors have in fact been found with quite a similar sesamoid, although he cannot speak to its function, of which he knows nothing, supposing simply that it was different from the present one. Yet a recent discovery in Spain, published in the review *PNAS*, on January 10, 2006, casts some light on the original purpose of this notorious thumb.

At Battalones, a site near Madrid, a Franco-Spanish team actually uncovered fossils of a precursor to the small panda, cousin to the large panda. Found in Southeast Asia, it resembles a raccoon and likes bamboo shoots, which it peels with its pseudo-thumb. It is believed that both had a common ancestor and diverged from one another several million years ago.

Back to our skeleton now, which has been named *Simocyon batalleri*.

This mid-sized mammal lived about 9 million years ago, and interestingly, also had a pseudo-thumb. Studies of its dentition, however, revealed that it ate mostly meat and not vegetation like present-day small pandas. It was therefore concluded that *Simocyon batalleri* did not use its sixth digit to grasp bamboo the way pandas do nowadays, but rather to help it make its way through the trees. French paleontologist Stéphane Peigné, member of the team, finds this a valuable asset in the particular environment, which contained a large number of predators:

> The odd precision with which Simocyon received its pseudo-thumb seems, under the circumstances, essential to this not-very-rapid scavenger, which needed the assistance in climbing trees.[21]

Though we may not be absolutely sure about the evolution of panda thumbs (*Simocyon* not being a direct forebear), we can say that the two pandas evolved convergently, apparently because they had the same ultra-specialized food, bamboo, which is extraordinary in itself. We can also say that the small panda is a wonderful example of what (at Gould's suggestion) is now called an "exaption" or secondary re-adaptation. In other words, it is a structure whose original function differed from its present one. Gould shows evolution working opportunistically from ready-made materials. The Spanish discovery provides an even more striking example of this fact. If Gould had not died in 2002, I'm sure he would have relished this!

There are still some things to be cleared up, however. Often science uncovers an answer only to reveal yet another question. Personally, I came across two that sent me back to the periodicals looking for more answers.

First, how much do we really know about the relation of our two modern pandas: the large one, *Ailuropoda melanoleuca*, and the small one, *Ailurus fulgens*? Despite their names, might they just be very distant cousins after all? I knew this had already been hotly debated, some likening the large panda to bears, and others to raccoons, while the small panda wavered between the two families depending on whom you read. In *PNAS*, Manuel Salesa and his colleagues sum up recent research based on molecular biology, and conclude with a phylogenetic schematic in the form of a tree that places them in two separate groups: the large panda (with bears, Ursidae) and the small with Ailuridae diverged from one another about 35 million years ago. Despite their names and appearances, they are phylogenetically quite remote from one another. Upon reflection though, this makes the convergent evolution of a common thumb even more intriguing.

Yet this sesamoid did not just appear all at once, as if by magic. This brings me to my second question: how, via what mechanism, was it made? Gould said nothing about it. Dwight Davis, a specialist at the Chicago Field Museum, however, wrote a detailed paper on the giant panda in 1964, in which he pointed out that the radial sesamoid and muscles surrounding it (abductor and adductor) were common traits. Both exist in all carnivores, but attached only at the base of the pollex, or true thumb. In common bears, the radial sesamoid is already more developed than in other carnivores. In fact, the abductor muscle ends in two tendons, one inserted in the base of the thumb, and the other in the sesamoid. Davis believes the entire string of muscle transformations springs automatically from a simple hypertrophy of the sesamoid bone. The muscles, then, would have mutated, because the enlargement of this bone did not allow them to attach at the same place as before.

Davis' analysis really becomes interesting when he identifies a possible genetic mechanism as the origin of this modification. It may all be the result, he maintains, of a simple genetic alteration, perhaps even a single mutation, which affected growth in the bone. He relies above all on the fact that the panda foot also has a sesamoid, likewise highly developed, though still not producing a new digit, and not conferring any advantage, despite its increased size. A mutation might have occurred, thinks Davis, causing enlargement of both sesamoids, tibial and radial.

This explanation would certainly be logical and correspond well with the manner in which vertebrate bodies are built. Since the 1980s, it is known that gene regulators (homeotic or *Hox* genes) exist and control development of the body in both space and time: body architects, if you will. More recently, the work of Denis Duboule in Geneva has shown that a single genetic control centre operates in the formation of digits (*Nature*, November 14, 2002). Subsequently, we can suppose that, in the course of evolution, one gene became inactive or was modulated by another, resulting in a larger sesamoid. All that remained was for the surrounding muscles to rearrange themselves accordingly.

Davis' hypothesis is very attractive and admits of later additions, as I have just shown, but it has yet to be proved. Likely, some researcher will provide a convincing solution to this enigma. Suffice it to say that one or two genetic alterations could lead to all these anatomical and functional consequences, including a very useful pseudo-thumb. I'd like to think that the same thing may have happened in primates a few million years ago, and been passed on to chimps and humans. Suppose chance had stumbled upon one or two decisive mutations in their most complex system of all — the brain. Presto, a set of physiological and anatomical changes gets underway, resulting in the big-headed *Homo sapiens*.

A Very Brainy Animal

IT IS OCTOBER 1860, within the panelled walls of Oxford University and the Royal Society for the Advancement of Science. A year after publication of *On the Origin of Species*, the interest of Victorian society is so intense that the large hall is packed to the rafters. Darwin's book said very little about the origins of humans, but it did drop unmistakeable hints that humans were descended from the apes, or rather an ape. Simply shocking! Among most thinking people, this idea radically contradicts the teachings of the Bible, and worse yet, looks down on these teachings. In the spirit of the times, humans can only have been created in God's image, distinct from all other animals.

The Oxford gathering has been called to debate this point, and a lively argument is expected, but Charles Darwin himself is not present. As usual, he has stayed behind in his country home, far from any polemic;

Thomas Huxley, the renowned biologist, will defend his ideas with his customary elegance and brio. Facing him is Archbishop Samuel Wilberforce, a man of considerable talents and an enthralling theologian and philosopher, ready to toss out any number of provocative points. "And you," he addresses Huxley as he closes his argument to great applause, "are you descended from monkeys on your grandfather's side or your grandmother's?" Huxley smiles, steps onto the dais and calmly delivers his presentation. In closing, he turns to the archbishop and addresses him personally: "To answer your question, let me say that, if I had to choose between an ape and a man who uses his personal influence to sway the public and who is unprepared to use logic in combatting the progress of ideas, I would doubtless choose the monkey." (Drawn in essence from numerous witness accounts at the time.)

One of Darwin's strengths was his ability to predict the place of modern humans among living things. From his theory, he formulated predictions that he himself was not able to confirm, because he lacked pre-human fossils, practically non-existent in his day. I consider this ability to predict important, because it is something we have a right to expect from a scientific theory, and a healthy, normal way of testing it: projections borne out by subsequent observation or experimentation. I emphasize this when debating with creationists, including the modern variety — upholders of Intelligent Design.

Well, what did Darwin predict? He readily saw monkeys as mammals that morphologically resemble us to a high degree. Seeing that humans have evolved just as every animal does, Darwin postulates the existence of a common ancestor shared with one of the large apes, or another now-extinct species. He predicts, therefore, that we would find fossils spread over time to show a progression toward species closer and closer to humans. He catches glimpses of a series of stages leading from modest-brained apes to more well-endowed ones, increasingly bipedal and capable of manufacturing tools. These ideas are not contained in *On the Origin of Species*, but Darwin does touch on them in his notebooks, as well as in *The Descent of Man and Selection in Relation to Sex*, published in 1871. He goes further, believing the chimp to be most similar to us, and Africa to be our common birthplace:

> In each great region of the world the living mammals are closely related to the extinct species of the same region. It is therefore probable that Africa was formerly inhabited by extinct apes closely allied to the gorilla and the chimpanzee; and as these two species are now man's nearest allies, it is somewhat more probable that our early progenitors lived on the African continent than elsewhere.[22]

Nearly 150 years later, such vision is worthy of applause. The discovery of numerous hominid fossils in Africa has lent magnificent support to Darwin's thesis. Similarly, genetics has in fact revealed the chimpanzee to be what Darwin called our closest "ally," with a mere 1 or 2 percent genetic difference between us. How can one not accept these discoveries as spectacular confirmation of his predictions? This is a demonstration that leaves even apologists for Intelligent Design in disarray. After demanding scientific procedures, how can they refuse the testing of these ideas? Once drawn to reason along these lines, they tend to feel trapped and look for a way out.

I.D.

Intelligent Design differs from creationism insofar as it no longer takes Biblical teachings literally. It maintains that the complexity of organisms can only be explained by a higher creative force, usually unnamed. The example of choice for I.D. supporters is bacterial flagella, a structure they cannot imagine emerging from successive evolutionary modifications. However, the idea that the complexity of an organism is proof of its creation by a higher intelligence is far from new. English theologian William Paley encapsulated it in 1802 by using the image of a watch: when we come upon a watch lying in a field, he says, its complexity convinces us that it could only have been made by human intellect, not by natural processes.

Rather than face up to this demonstration (accessible to anyone with a grounding in anthropology or biology), they opt for a highly technical discussion of bacterial flagella — their favourite example of a highly complex structure which, they maintain, can only have been conceived by an intelligent force. They appear to adopt a strategy of piling up objections on points of detail, employing highly abstruse arguments that only the most fastidious specialists can disentangle.

Personally, I have no intention of embarking on a discussion of flagella biochemistry, a subject I know nothing about. Yet one has to reply to detractors of the theory of evolution by natural selection. Their position is quite simply wrong and undermines society's trust in science. I therefore have no hesitation whatever in discussing the predictions Darwin made about the origins of humans, and I highly recommend the exercise, striking as the example is. Here, surely, we are entitled to say, "Bravo, Mr. Darwin."

TESTING

This concludes a brief discussion I find useful in these times of intellectual confusion. Testing predictions is a hallmark of science. *On the Origin of Species* in 1859 convinced the scientific world that evolution by natural selection is a better explanation than William Paley's argument for Intelligent Design, because it could be tested and eventually proved or disproved. The argument for Intelligent Design, however, cannot be tested, nor can creationism: thus they are not science.

It's always fascinating to come face-to-face with a chimp in a zoo: the penetrating gaze, hands so similar to ours, the half-stooped way of moving around, the familiar way of scratching head and torso, the mimicry of an expressive face. We see a reflection of ourselves, a window into our own animal nature. A friend of mine claims this fascination is artificial, merely part of a culture that informs us we are close relations. What matters to him would be to know what the chimp is thinking and whether he sees a similarity in us. I can't agree. I think the interest most humans have for this animal is both spontaneous and profound. Experiments in psychology show that, regardless of culture, young children always prefer the great apes. Even Inuit babies not at all familiar with monkeys are attracted to pictures of them as much as to polar bears or dogs.

Watching a group of chimps together, we spontaneously interpret their behaviour in terms of our own — an attitude we adopt toward other animals, but less so. We look for meaning in eyes so similar, a way to communicate, to read emotions. Despite this, we can still feel a certain fear of the wild and unpredictable animals we see in them, and not without reason, for a gorilla, an orangutan or an adult chimpanzee can easily bring down even the toughest of humans. One anecdote puts this in perspective: In April 2006, in a Sierra Leone wilderness sanctuary, a taxi driver and three Americans working on a nearby site were attacked by chimpanzees. The taxi driver was killed, and the Americans spent two months in hospital.

Although zoologists and anthropologists have tried to link us more closely to the gorilla or orangutan in the past, we now know we are nearest to the chimps. Orangutans separated from the other higher primates 19 million years ago, and gorillas branched off 8 or 9 million years ago, whereas chimpanzees and bonobos formed the branch from which humans would emerge only 5 or 6 million years ago. (Things then get a bit complicated, but the general picture of an evolutionary bush spreading far and wide, with some dead branches, so as to result eventually and quite recently — about 200,000 years ago — in *Homo sapiens*, our ancestor, is a reliable one.)

The fact that modern biology places humans and chimps closer to one another than we thought two centuries ago is still rather disturbing for a number of our contemporaries, even 150 years after Darwin. Most telling are the "killer details," the percentages that measure this proximity. Generally, the genetic resemblance is considered to be 98.8 to 99 percent. These figures, and their inverse, affirming that we're different by a mere 1 percent, need to be put in context. Let's hit "pause" for a moment and see just what modern genetics tell us about this.

The first comparisons of DNA in primates in the 1980s used a somewhat rudimentary technique called DNA hybridization, which amounts to comparing reference points on the overall genome. It allows for approximate estimates which led us to conclude that men and chimpanzees shared 99 percent of their DNA, men and gorillas 97 percent, men and orangutans 96 percent.

Fifteen years later, when genomes were being sequenced, comparisons became more exact. (Sequencing is the process by which the order of amino acids is determined in proteins, or the order of nucleotides in corresponding portions of DNA.) Researchers noticed that humans do in fact differ very little from their closest animal relation, the chimp — 1.2 percent in the bases that can be aligned together (compared exactly between the two), thus confirming earlier data. The difference between corresponding proteins is in the order of two amino acids, on average, per protein. However we look at it, the chimp remains our closest cousin. We definitely had a common ancestor about 6 million years ago, and our genes are there to tell the story.

Still, there is a striking paradox in this closeness; although we resemble *Pan troglodytes* tremendously, we remain very different just the same. *Homo sapiens* lives far longer, is bigger, stands fully upright, is less hairy, and has shorter arms, with a larger pelvis, a smaller jaw, and a brain four times bigger. These are not mere details. We alone possess language and handle words and figures, follow complex religious rites, cook chow mein and osso bucco — complicated dishes whose geographical and historical origins matter to us. We alone ask questions about our relation to chimpanzees. How does such a minimal genetic difference allow for all this? Do we have "uniquely human" genes, and what do they actually do?

Nowadays, not many zoos keep chimpanzees in captivity. They are protected in the wild, and the ethical debate is intense, opening both minds and cages wide. It's probably better that way too; let's just protect them and watch them on TV, thanks to the wonderful Jane Goodall. If you've had the chance, as I have, to see them in a zoo where they are well treated, you may have seen something like this: it was mealtime, and the chimp attacked his meal, his powerful jaws grinding the green shoots of bushes, extracting the essences and leaves and chewing them for a long time — fascinating!

The strength of those jaws is part of what distinguishes them from us. One of the muscles that close the mandible is far more powerful in them. (The temporal is the one we feel stretching across our temples, just above and in front of the ear, when we close our mouths slowly then relax the pressure.) In the chimpanzee, the cranial surface it attaches to is much greater.

Why is this detail so important? It happens that in 2004, a team led by Hansell Stedman of the University of Pennsylvania noticed that the gene responsible for strength in the temporal muscle (MYH16) had undergone mutation in humans, relative to chimps, gorillas and macaques. Its strength depends on the myosin heavy chain 16 protein, and there exists a mutation that prevents this protein from forming, with the result that the fibres of human temporal muscle are reduced in size. Stedman managed to show that deactivation of the MYH16 gene in humans occurred through evolution about 2.7 to 2.1 million years ago, a little before *Homo habilis*, but clearly around the time the *Homo* genus began to appear. Knowing that this period corresponded to a phenomenal increase in the size of hominid brains, one is tempted to see MYH16 as primordial to the difference between men and chimps.

As we know, *Homo sapiens* is distinguished by having an even larger brain than *Homo habilis* as well as a smaller jaw: was the mutation of the gene synthesizing the myosin heavy chain 16 protein a decisive stage in this journey? It may be too soon to tell, since we need to map the entire route first, but it is possible that shrinking the jaw allowed the brain box to enlarge, or vice-versa: was it the result of other anatomical alterations that, in turn, were controlled by genes with an even more decisive role in evolution?

The fact remains that this gene is the second involved in the evolutionary distinction between humans and chimpanzees, another having been discovered a few years earlier. The story is worth a brief recounting, even though, as with MYH16, it is difficult to draw firm conclusions.

In the early 1990s, English researchers were intrigued by an odd language impediment suffered by a family in suburban London. Half the family (twenty out of forty over three generations) had difficulty articulating and understanding syntax and grammar. It took great effort for them to be able to speak by the time they were adults. Could it be they were subject to the mutation of a gene crucial to language? Since it all occurred within one family, it seemed to be a hereditary trait, but it took eight years to pin down the gene responsible, called FOXP2.

What happens next is fascinating. Studies have shown that all humans who speak normally have the same version of FOXP2, no matter what culture or latitude they come from, but not this British family, *nor any of the higher primates*. Chimpanzees, for example have no "human" FOXP2, but rather a variant with only two base differences. We still don't know precisely what this gene does, but we do know it expresses itself in the brain and has a "domino effect" on other genes. So it is a gene regulator. Since there are so few genetic differences between the two species, it is hardly surprising to stumble on a gene like this.

Other distinctive genes will surely be found in coming years, and there are obviously some in the reproductive and immune systems, chimps being affected quite differently by viruses and bacteria. They are resistant to malaria, for example, whereas we are highly sensitive to it, but they are more vulnerable to tuberculosis. We can expect their olfactory genes to be different or more numerous, since their sense of smell is more highly developed than ours, but the greatest difference has to be in the brain. What evolutionary path was it that allowed us to be so advantaged in that area?

Here, science is more in its infancy, so consensus is not yet possible. Some believe we will soon identify a series of genes to explain this difference, the key residing in their expression. One example might be the ASPM protein, at least partly responsible for brain size; another might be a gene discovered in 2005, and expressed far more in the area of human brains involved in higher reasoning than it is in chimps. These isolated breakthroughs do not provide an overview, but the hypothesis remains that a fairly small number of genetic mutations in the two species could explain everything, that is, if they are gene regulators involved in brain development — the famous architect genes.

Still, this hypothesis is not universally popular. The neurobiologist Robert Sapolsky at Stanford University, in particular, defends the alternative theory, that the answer to the 1.2 percent paradox lies not in the existence of different genes, or differently expressed genes, but simply in the total number of neurons. In the April 2006 issue of *Discover*, he begins by recalling that, despite the difference in brain size, humans have very few unique characteristics.

The chimp, for example, has function zones equivalent to our language areas, the Broca and Wernicke zones. These are asymmetrical in human brains, as their counterparts are in chimpanzees. Furthermore, both species have the same neurons, the same neuro-transmitters, and the same mechanisms for circulating them. Here, though, comes the difference, and what a difference it is! Humans have far more neurons, which appear at the embryonic stage with the inception of the brain.

At first, Sapolsky explains, all embryos possess an initial cell, which will split to create others: 4, 8, 16, and so on. He offers the following:

> After a dozen rounds of cell division, you've got roughly enough neurons to run a slug. Go another 25 rounds or so and you've got a human brain. Stop a couple of rounds short of that and, at about one-third the size of a human brain, you've got one for a chimp. Vastly different outcomes, but relatively few genes regulate the number of rounds of cell division in the nervous system before calling a halt. And it's precisely some of those genes, the ones involved in neural development, that appear on the list of differences between the chimp and human genomes. That's it; that's the 2% solution. What's shocking is the simplicity of it.[23]

Essentially, Sapolsky's hypothesis is not so far from the architect genes; rather, it is just a variant of it, holding that genes controlling the synthesis of neurons in the brain play the all-important role of regulator genes. I applaud Sapolsky's enthusiasm and appreciate the brilliance of his solution, but he offers no demonstration, not even the shadow of proof. It is a very attractive hypothesis, but just that.

It does show, I believe, that we are at the very beginning of exploration into the genetic basis of chimp-human differences. Science is just starting to sequence the two species, noting differences, mysterious duplications of bits of DNA (more numerous in humans than in other primates). Occasionally, by luck, as with MYH16 and FOXP2, researchers stumble upon a particular gene. This is exciting, but we're still a long way from having fully inventoried the differences and the way in which they appeared over 12 million years of evolution.

Perhaps it would be better not to look for "distinctive genes" as the last word. Studying the differences between the species and within humans shows that evolutionary change happens very slowly and through numerous genes, each one responsible for an accumulation of small effects. There is also good reason to believe that numerical differences in genes are not the most important. Finally, as we have seen in preceding chapters, the "basic kit" of genes is similar among all mammals, the difference mainly being in how they are used. The image mentioned by Sean B. Carroll is the one that recurs most frequently for me; it is a nod to guitarist Eric Clapton — "It's all in how you use it." All guitar players would probably agree: it's not the instrument that counts, my friend, but your fingers on its neck!

In the end, all this perhaps allows us to solve the 1.2 percent paradox as well as the diversity of all living things. Both phenomena appear to have a common origin, and evo-devo would seem to hold the answer: small modifications in regulator genes can rapidly increase the complexity of systems and contribute significantly to biodiversity. That's it! Excuse me, but like Sapolsky, I've just had a flash. That's indeed it — the answer to the 1.2 percent paradox. Simply shocking!

PART THREE

When Humans Interfere with Evolution

Parasites in High Gear

LET'S DO A BRIEF recap, a bit like those detectives in novels just before they explain the outcome to us. Biological evolution is driven by two main factors: the *variability* within populations (due to random genetic mutations) and *natural selection*, which Darwin referred to as "descent with modification," called "reproduction differential" by modern biologists. These two agents are continuously at work over deep time and in millions upon millions of individuals. They work blindly on basic materials already at hand, in what François Jacob calls "the play of possibilities." Some survive and reproduce, while others do not. New forms appear — and we've seen how genes are important in embryonic development — so once in a while, some innovation will emerge. If it aids individual adaptation to the environment, it will remain; if not, it will disappear. Humans are neither a culmination nor a planned result of all this. Like any other species, we are the random product of chance.

It is nevertheless a fact that no other species has been as hard on its environment or on other species as we have. In hunting and fishing species, in domesticating them along with plants, in expanding our population over the Earth, we have created upheavals hitherto unknown in nature. Of course, the history of life on this planet shows tremendous disturbances, but all of them until now are from purely natural causes. Paleontologists have, for example, found rock layers showing total and rapid die-outs of species ("massive extinctions"), followed by reconstruction. There appear to have been five of them since life began. Experts attribute this to internal causes such as intense volcanic activity or external causes such as meteorite impacts. The fifth massive extinction of the dinosaurs about 65 million years ago was probably due to a giant meteorite, which left traces of its crash in the Yucatan.

The extinction of species has been going on so rapidly over the past few thousand years that paleontologists have no hesitation in talking about an impending sixth massive extinction, this time caused by humans. Niles Eldredge of the U.S. has traced the start of an astonishing acceleration in extinctions to 30,000 years before the modern age. He points out that this coincides with massive human hunting of a hugely rich prehistoric fauna, which then progressively dies out, especially in Europe and Australia. With the development of agriculture and husbandry about 10,000 years ago, another step is taken, and many wild plants disappear, while domesticated mammals severely disturb ecosystems and accelerate deforestation. One has only to recall that ancient Greece was covered with forests.

After the beginning of human settlement, the exponential demographic rise is rapid. Simply put, about 10,000 years ago, there were only about 100,000 humans, 1 billion at 150 years ago, and more than 6 billion today. Nature is frankly under attack, as plants are harvested, eradicated or deprived of habitat. Except for the poles (with little life anyway), no nook or cranny is safe from human domination. The International Union for Conservation of Nature estimates we have wiped out 800 species of mammals alone in the past few centuries, and that is one-fifth of those remaining. Intensive hunting and overfishing now threaten an unprecedented number of species.

The exact magnitude of the worldwide biodiversity crisis is, of course, a matter for debate beyond the scope of this book. Past extinctions have taken hundreds of thousands of years, however. It might be argued that we are still too close to the problem, but it is clear at present that our actions are tantamount to a natural cataclysm. It is quite something to admit that humanity's ravages are equivalent to a century of continuous intense volcanic action or the crash of a giant meteorite!

What does it mean for evolution when a single species possesses as much power as we do? Humans don't have any direct power over natural selection, nor do we influence genes (although in some cases we have begun to — we'll return to this in Chapter 11). Yet simply by the disappearance of many animal and plant species, by the speeding up or slowing down of expansion in others, we instantly affect evolution. Two of the most striking examples are overfishing and the evolving resistance to bacteria and pathogenic viruses. To put this into a better context, we need to enter into the microscopic world of co-evolution and parasitism.

Parasitism is a fundamental dimension of life. In fact, we know of hardly any organisms that don't have parasites, however slight their complexity might be. A considerable number of viruses, bacteria and insect species can survive only inside host organisms, or in symbiosis with them. They hang on, diverting food and shelter for themselves. The relation might be one-way (simple parasitism), or it might serve both (symbiosis or mutuality), and it touches all realms of life, leading to amazing alliances, say, between ruminants and birds who rid them of stinging insects. Sometimes it occurs within a single species, with some tiny male fish clinging to the sides of females of the same species to collect leftover food. Cuckoos, of course, use guile, laying eggs in the nests of nearby birds belonging to another species. Once hatched, the newborn then quickly expels the "legitimate" chicks and tricks the parents into letting it monopolize the food.

Generally speaking, says Claude Combes in *Les associations du vivant, L'art d'être parasite* (*Life and Its Associations: the Art of Being a Parasite*), parasitism becomes an evolutionary path maintained by the host species, so each parasite has to find its partner. That is its mission and an absolute

STRATEGY OF THE SMALL LIVER FLUKE

The strategies found in the world of parasites are weird and wonderful. The small liver fluke, for instance, is a trematode that attaches itself to sheep and cows. To begin with, eggs are excreted in the infected animal's feces, then eaten by certain types of ants. How, then, are they to make their way back to the host animal? Well, a ruse is called for, and the small liver fluke influences its host ant by getting it to climb up a blade of grass and then waits to be consumed with it by the grazing animal. More precisely, its larva lodges itself in the ant's nervous system, thus changing its behaviour. Once in the larger animal's stomach, a nice adult sheep, let's say, it moves on to the liver and begins its attack. Then, it returns to the stomach to lay hundreds of eggs and die, and the cycle continues.

necessity: no host, no life, no reproduction. The host, however, must constantly struggle to avoid this partnership or at least diminish the harmful effects. Each species then selects genes through evolution (especially immune defence genes) that will either help avoid the invader, neutralize it, or better yet, kill it. The parasite, for its part, furbishes its arms by degrees. Natural selection allows it to favour genes that will lead it to a partner, then help enter its body or attach to it — a regular arms race!

Not taking into account the genetic element, about which he knew nothing, of course, Charles Darwin had already described these phenomena very well. He referred to them as *co-adaptation*, whereas contemporary biologists prefer *co-evolution*. Are they incidental without profound significance, or do they represent a fundamental evolutionary force? The latter was long preferred, co-evolution being considered an illustration of

Alice and the Red Queen running hand in hand. Illustration to Through the Looking Glass *by John Tenniel. Wood-engraving by the Dalziels.*

the adaptive ability of living organisms. However, in 1973, a researcher at the University of Chicago, Leigh Van Valen, re-launched the debate when he advanced what is now called "the Red Queen Hypothesis."

The expression of course comes from Lewis Carroll's *Alice Through the Looking-Glass*, in which Alice and the Red Queen hold hands as they run through Wonderland. After a time Alice notices that for all their effort, the landscape around them hasn't changed. Asking the Red Queen about this, Alice is informed that they are running to keep themselves where they are, to which she replies — somewhat out of breath — that where she comes from, they would normally get somewhere, somewhere else. The Red Queen finds this astonishingly slow and says, for them actually to get anywhere, they would have to run at least twice as fast.

Van Valen finds this a good comparison to what occurs in co-evolution: species in conflict run — that is, they keep inventing new adaptations, but in this "arms race," they simply maintain equilibrium. Widening his hypothesis to the predator-prey relationship, he postulates that competing species are actually *hostages* to an unending evolutionary race. Evolution allows predators to become better at finding and catching their prey, who in turn grow better at evading them. Both improve their performance, so their relative numbers do not change, except perhaps in the short term.

The arms race between parasite and host casts new light on humans. Strictly speaking, we do pay a price for the parasites that feed off us. These are tiny creatures like worms and even single-cell creatures like amoebas and paramecia, which can cause terrible illnesses. Paludism (malaria) and schistosomiasis alone kill over 2 million a year.

This race becomes even more deadly when it comes to bacteria and pathogenic viruses. Some bacteria can rapidly inundate the human immune system. Viruses, on the other hand, cannot multiply by division and depend entirely on a host cell. These are *obligatory parasites*.

Since these foes were first identified barely two centuries ago, we have trained our medical arsenal on them. After a fairly confused beginning full of trial-and-error, our arms are now well furbished.

The discovery of penicillin during World War Two was a major breakthrough against bacteria, and pharmaceutical companies promptly perfected a series of antibiotics. After that first golden period, however, the sheen began to wear off. Our foes defended themselves by developing resistance to the arms we lobbed at them. This resistance has now become a frequent occurrence and applies itself to multiple antibiotics at once. Fifteen years ago, certain strains of Streptococcus A required 10,000 units of penicillin; now they require 24 million units. Even this is sometimes not enough, and death is the result. *Staphylococcus aureus*, the plague of post-operative infections, has become resistant first to penicillin, then to streptomycin, tetracycline, methicillin, and even at times, vancomycin — until recently, the antibiotic of last resort.

What on earth happened? Again evolution can explain this important phenomenon, hitherto unknown because it was caused by modern humans in a very short time. With the first mechanism, resistance occurs to antibiotics that are too weak or taken in insufficient doses, and they thus kill off only the weakest bacteria; natural selection allows the strongest bacteria to survive and spread more than ever before. With the second, resistance can occur as a response to the antibiotic; mutation causes the bacteria to modify their structure. As a result, medication can no longer penetrate the bacteria's membranes. A third phenomenon might be added to this, one whose importance is only just beginning to dawn on us: resistance is amplified by the amazing capacity of viruses and bacteria to swap genes on contact (*horizon-*

tal genetic transfer, since it is non-genealogical). This occurs between bacteria or from virus to bacteria.

When it comes to viruses, we know antibiotics don't work, but recently we have come up with effective antiviral medicines. As a general rule, it is the characteristics of the virus itself that limit the strategies for fighting infections. It is hard to eliminate them without killing the host cell, because they use its own machinery to reproduce from inside it. One possible approach is to vaccinate, but this is not always possible, and it still remains to defeat the virus' ability to mutate rapidly. HIV is the best example of this. Someone infected by the virus turns into a veritable walking evolutionary lab: the virus continually adapts to the host's immune system and to the medications used.

This may all seem very disheartening, but as we explore further evolutionary avenues, some may be particularly fruitful. In his book *The Evolution Explosion*, biologist Stephen Palumbi of Harvard offers some interesting clues:

> Drugs are evaluated on their potential to kill viruses. Fair enough. But the virus is not the only enemy we face. Another foe is the evolution of the virus, and few drugs are evaluated on their ability to kill this process. Furthermore, if drug resistance is inevitable, then by choosing drugs, we are in effect choosing the evolutionary trajectory of the virus. Why not use this opportunity to channel the virus into an evolutionary cul-de-sac and then let loose the pharmaceutical dogs?[24]

Summing up, he writes: ". . . while HIV evolves, it thwarts us. And so we must learn how to control evolution in order to survive the evolutionary skills of HIV."[25]

What an original idea, and one I believe needs further exploration. When are we going to see an international task force of evolutionary specialists look for a strategy that will force HIV to beat a retreat? If we have managed to accelerate the evolution of bacterial and viral parasites, perhaps we can make up for it by wresting back control over their evolution!

Salmon under the Influence

MY MEMORY OF THE Atlantic salmon is utterly breathtaking: September at the edge of a river of creamy water in the Gaspé, cool following the rain. Three fishermen in hip boots stood at a crossing-point that bordered a salmon pool (one of those basins where the fish rest on their journey upriver), and three canes whipped the air so elegantly that their flies barely stroked the water's surface. In vain, it seems — for an hour I watched them, seated on the rocks, lines occasionally stretched taut, nothing but false alarms.

My fixed concentration on the rock at my feet eventually yielded an occasional longish shadow which might have been an Atlantic salmon, but I couldn't be sure. Lower on the river, a wooden salmon ladder had been set up and was spilling yellow water, and there I could see them up close. These adult fish were ascending the river they were born in, some for the first time in their lives, and others for the second or third, which is possible, at least for this species. Built for speed and for battle, they cast long, silvery shadows 50 to 90 centimetres long, making their way up each step of the ladder with a vigorous tail stroke, their entire bodies arched in motion — superb animals, wonderfully adapted to this difficult task, working against the strong current and the length of the river.

How to express my admiration for salmon? I love how they look, their physical strength and their diversity, particularly the two genuses *Salmo*, which gave rise to the Atlantic salmon, and *Oncorhynchus*, best known for its seven Pacific salmon species.[26] I'm drawn to their various complex life cycles, from freshwater birth to whitewater adolescence, then adult life in the sea, and finally their legendary struggle to return, spawn and die in the river of their birth — a spectacular and romantic journey.

In the fall of 2006, while filming a piece in British Columbia for *Découverte* on Radio-Canada (the French counterpart to the CBC), I witnessed sockeye salmon returning to spawn in their native river, and what a show it was! That year, in the Adams River Basin, 2.5 million of them had started out from the ocean, but as usual, most of them died along the way. They had to face the currents, the stress of the journey and parasites, along with commercial, sport and traditional native fishing in the straits and on the lower Fraser River.

The fish no longer feed once they reach the river, but just live on their reserves. The males grow a kind of snout, which helps in their battles for females. The meat of the sockeye becomes redder and redder as the journey progresses because of the reduction in fat, which leaves the skin transparent and the muscles reddish — not worth eating, because it lacks moisture and flavour.

Finally, in a last effort, the luckier ones reach the point where the Adams River flows into Shuswap Lake. (This too is an intriguing mystery, because they seem to remember and locate their birthplace through the unique smell of the water.) In this channel, the current is very strong, and many salmon fail this final test, but if they succeed, they are rewarded about 300 metres upstream with a calmer, wider river. Now the mating dance can begin, and the water is so crowded that there are flashes of red everywhere; fights between males cause couples to be formed, broken and reformed, the females then laying thousands of eggs at a time. The males cover them at once with milt, and new lives are underway. Inevitably they mate and die within a few days. Then, in spring, the eggs hatch, and a new four-year cycle begins.

Salmon have a complicated evolutionary story. Long before humans, about 50 million years ago in the Eocene Epoch, lived the oldest salmon ever to leave us its fossil. The *Eosalmo driftwoodensis* was discovered by a University of Alberta paleontologist in 1977 in the sediment of Driftwood Creek near Smithers, British Columbia. Only about 30 centimetres long, the specimens' sleek outline, common to later salmon, trout, arctic char and whitefish, is clear. The sediment is typical of freshwater lakes and rivers.

The presence of specimens from all age groups indicates that this ancestor did not go down to the sea to grow into an adult. In the first half of the Miocene Epoch, 15 to 20 million years ago, two major genuses emerged: *Salmo*, which would later colonize the North Atlantic — thus also the rivers of eastern North America and Europe — with *Salmo salar*; and *Oncorhynchus* in the North Pacific and rivers emptying into it, both east and west (seven species: sockeye, chinook, pink salmon, chum or keta, coho, masu and amago — the latter two only on the Asian side).[26] Fossils tell us this flourishing of Pacific species occurred from the mid-Miocene about 13 million years ago, to the beginning of the Pliocene about 2 million years ago.

What these fossils can't tell us, genetic analysis can: in other words, 25 to 100 million years ago (sorry for the vagueness!), a genetic slip-up turned salmonids into tetraploids. Instead of having a pair of each chromosomes like most animals, they have four. As a rule, this type of mutation proves fatal not long after the animal's birth. In a few rare cases like this one, it does not affect survival, but merely gives the animal a double set of genes. The advantage is that new mutations can appear in the duplicate chromosomes without much risk, since the organism's proper functioning is guaranteed by the "backup." What a lab this provides for natural selection!

Frequently, species that become tetraploids are a little more rugged than their normal counterparts. This advantage might have influenced the evolutionary success of the many species of Pacific salmon. Still, the *Salmo* genus had it too, so why did it yield only a single species? David Montgomery of Washington State University at Seattle has a convincing hypothesis.

In a 2000 article, he begins by stressing that the period in which Pacific salmon so broadly expanded coincided with significant geological shifts in western North America. The Alaska Chain down the coast of British Columbia, as well as the Cascades and others farther south, all came into being at about this time in response to intense volcanic activity in the region. The Atlantic coast on the other hand, saw the Appalachians formed much earlier, then geological stabilization around 70 million years ago, with little disturbance from migrating glaciers and rising sea levels. So diametrically opposed are these two geological regions that they might explain the different fates that befell the *Salmo* and *Oncorhynchus* genuses. It is possible, Montgomery says, for the geological "churning-up" in western North America to have created natural barriers, changing the directions of waterways and perhaps isolating some populations enough to create new species. This speciation is, in fact, not yet over, since some Pacific species can produce viable hybrids when they intermingle.

Coho, sockeye and chum: these are the best-known types of canned salmon on supermarket shelves. Nowadays, however, like their Atlantic cousins, they are in trouble, as their numbers fall relentlessly. Overfishing is most certainly one culprit, but climate change is another likely cause.

The sockeye, for instance, spends most of its life in the ocean. In the past ten or fifteen years, however, greater quantities of fresh water have been noticed in the Georgia Strait just off Vancouver. This appears to be the result of runoff from glacial melting, a result of global warming. In reaction to it, the salmon may, on their way back, adapt more rapidly to life in fresh water, hence making their way upriver earlier on, but too soon for the amount of energy they have. This would then influence their reproductive capability. This hypothesis has yet to be confirmed, but researchers involved with the question are concerned about the species' survival, at least in the Fraser River. The fact that the September 2010 Fraser River sockeye run was unexpectedly enormous (some 34 million fish) after the disastrously low runs in the previous three years is even more cause for concern since it shows that the federal Department of Fisheries has little understanding of the situation.

The amount of fishing, of course, has a direct impact on the evolution of salmon, as it does on other heavily-fished species, like tuna and cod. Natural selection favours the larger females, which produce the most eggs, but these are precisely the ones most affected by intensive fishing in oceans and coastal areas. This results in those left to mate in rivers being often the smallest and the slowest-growing. In normal circumstances, the latter would have been eliminated by natural selection, but now they survive and reproduce, thanks to the big, check-shirted predators who are on the prowl with their nets.

Since such selection repeats itself from generation to generation, the average size of salmon is shrinking. A study by W. E. Ricket shows the average size of two-year-old pink salmon fished at Bella Coola, B.C., went down 30 percent from 1951 to 1974, and that is huge. Is it mostly because the larger fish are eliminated by ocean fishing and never make it to their spawning grounds? Although it is certainly a factor, the answer must be no, since all the salmon studied were of the same age, the age of the spawning run. The decline in average size then clearly shows an evolutionary tendency.

The salmon have also taught us a great deal about what scientists call "contemporary or rapid evolution," taking place over a very short time on the geological scale. It is so short, in fact, that we can directly observe it over decades or centuries. We are talking here of complex organisms such as vertebrates, not mere bacteria or viruses capable of evolving quickly and reproducing in a matter of minutes or hours. Surprisingly, animals that take years to reproduce can evolve in mere centuries or even decades. This is a major revelation for specialists, and one that neither Darwin nor his immediate successors foresaw. It is worth noting that the most rapid evolution occurs under the influence of humans.

We have just referred to the evolution toward smaller sizes of fish caused by overfishing, and this is generally accepted now by researchers, but it has only come to the fore in the past fifty years. There are other more recent discoveries.

Back to the beginning for a moment: a fundamental question of evolution is how species are formed. We know the general answer given by Darwin: most often, one species yields another in smaller, more marginal populations that become cut off from the original population (specialists call this *allopatric* or *geographic speciation*). In reality, though, it is less simple, and speciation is almost never directly observable, so there is considerable debate which we will not delve into here, but suffice it to mention the discoveries made in salmon.

In 2000, scientists measured the rate at which a founding population of salmon in two distinct ecosystems can give way to two populations no longer able to reproduce together (a stage in speciation). Everyone had been convinced it took centuries or even millennia, but Andrew Hendry at the University of Washington — now at McGill — has found that it requires only thirteen generations, or barely sixty years.

Here is the context: salmon had been introduced into Lake Washington between 1937 and 1945, and the genes of two distinct offshoots of the original population were studied — one group spawning along a beach of the lake, and the other in one of its tributaries. Using DNA, Hendry and his team found proof that, despite the accessibility of one territory to the other, there was no longer any genetic exchange whatsoever. Even if the river salmon were sometimes present at the beach, there was no crossbreeding. The conclusion clearly is that the two populations had become entirely separate species; after only thirteen generations, speciation had already occurred!

A further study by biologists at the universities of British Columbia and Windsor has fuelled the debate over what can be called accelerated evolution. Their 2003 publication shows a rapid change in the size of salmon eggs in certain populations. Here again, some detail is necessary to appreciate the breakthrough. A common practice in wildlife management is to capture adults swimming upstream to mate so as to have the eggs fertilized in a controlled environment. The newborn are then raised in captivity and finally released into the wild to boost declining populations. This is where things become difficult: their first "comfortable" months favour traits suited to captivity, not life in nature. Chinook salmon in particular experience less of the natural pressure to produce large eggs, since the young are well nourished from birth and survive without any problems. Females thus tend to produce mostly small eggs. Evolution has thus "deviated" toward smaller and smaller eggs, which in nature is a dangerous situation, since the fry need at least a minimum of reserves at the beginning of their lives. This was observed by the Canadian researchers over a period of fifteen years.

Charles Darwin, believing that evolution occurred only gradually over thousands of years, would probably be highly surprised at such results. It is known that he defended a theory known as *gradualism* for the steady, regular and slow way in which species were supposed to transform themselves: *Natura saltum non facit*, or nature makes no leaps. More recently in the 1970s, Americans Niles Eldredge and Stephen J. Gould proposed the alternative *theory of punctuated equilibrium*.

This theory holds that, in the relation between a species and its milieu, evolution alternates between long periods of balance and short ones of abrupt change. Most of the time is taken up with prolonged periods during which life remains stable, as fossils do indeed show. Then significant modifications occur very rapidly. Yet Darwin and followers of the synthetic theory expected the appearance of a new species to take millions of years. Eldredge and Gould maintain that it takes only a few thousand years. They showed that the emergence of a species occurs through a millennial accumulation of genetic mutations and morphological alterations in a small population subject to different conditions from the rest of the species. The small size of the population would then account for the absence of fossils at various intermediary stages. After much debate, it can be said that the theory of punctuated equilibrium is generally accepted by evolutionists. It does not fundamentally challenge Darwinism, since it is concerned only with the pace of evolution.

In this debate, evolution is always viewed in the very long term, without any special consideration for the recent arrival of humans, for we are merely a brief twinkle in the story of living things. Nevertheless, the evolution of salmon in response to human pressure is so rapid that even followers of the "nature progressing by leaps and bounds" idea could never have imagined such a result. Humans as the accelerators of evolution? Indeed, and perhaps even more than we think, and with more serious implications than we suspect.

Mr O'Connor's Stubborn Struggle

O'CONNOR'S FARM IS A mere two kilometres from our cottage in the Eastern Townships, but it might as well be a world away. On a hillock by the edge of the forest, it faces south toward a small, wide valley which has been completely deforested and combines lovely pastures with corn and canola fields. In the other three directions are wooded, rocky fields and some watering holes with bushes that seem to suit Mr. O'Connor's herd of cows just fine.

At summer's end, we go over to buy vegetables and freshly picked corn. It's all delicious, inexpensive, and Mr. O'Connor (I've never known his first name) loves to chat. He's over seventy if he's a day, and he can do as he pleases now his sons look after the farm. A third- or fourth-generation anglophone, he speaks French with a strong accent and mixes the two languages with joyful abandon, but he manages. Picking out the vegetables and weighing them on a rusty scale in his shed, old farmer that he is, he chats about the weather these past few days and the ones to come. His real passion though is the land, its quality and how it affects what he grows.

On this, he is inexhaustible: "Here there's tons of rocks," he likes to say, "but once you've got them out, the earth underneath is good, and we get the right amount of rain. Up on the hill we get fine grain and feed for the livestock." Most of his corn, except for a little he keeps to eat himself or sell to seasonal passersby, he keeps for his cows. He's also got chickens and rabbits, as well as a few beehives. He makes a point of telling me he grows "ordinary corn, not that modern stuff full of GMOs [genetically modified organisms] — nothing but trouble for animal and human health." At this point, his bushy eyebrows are definitely up. He holds it against his neighbours lower down in the valley that they have planted genetically modified corn these past years. In this instance, they farm a variety with a gene of the bacterium known as *Bacillus thuringiensis* (B.t.). It is a frequent occurrence, and a Université Laval study in 2003 found that nearly 80 percent of Quebec corn producers have used GMOs on their land. Mr. O'Connor is therefore part of a holdout minority, labelled "non-users" in the study. "Users," on the other hand, are also referred to as "the vanguard," oddly unscientific as a choice of words.

The gene of the B.t. bacterium in such corn controls the manufacture of a protein which is toxic to insects. The target here is the corn borer, a common pest in cornstalks. Working in all cells of the plant, the toxin kills larvae by perforating their digestive tubes immediately upon consumption. The larvae literally explode from the impact of this internal bomb, their stomachs peppered with a thousand holes. The beauty of this, according to its promoters, is that B.t. is harmless to humans, since our stomach acid destroys it (and it must be admitted, a score of studies do bear this out).

Theoretically, B.t. avoids the need to spread insecticide, which should reduce costs and pollution, as well as perhaps increasing yield. Genetically modified corn has been approved by the authorities and marketed in Canada since 1996. So why is Mr. O'Connor annoyed with his neighbours? He has a vision of his own, one that extends to the idea, "When you're a farmer, you've got to think long-term about things like people and animals and plants, because it's all connected. Simple as that!" The neighbours' fields completely surround his, and already he is concerned his crops will be affected by their weed killers; most of all, he is worried about their GMOs.

"In spring, all the pollen from their GMO corn is carried onto every plant I own," he says, "so what do you think my cows and rabbits are grazing on, and my bees pollinating? They take in all kinds of B.t. pollen; it's toxic for them and us. That's not all. It doesn't increase yields, and it doesn't reduce the need for herbicides and insecticides, but it does poison our health. It's already proven to harm bees; I can see that with my hives. It poisons the birds too, the mice and farm animals, let me tell you! One of these days, researchers are going to find out the B.t. residue in our food is destroying helpful bacteria in our stomachs and causing sicknesses that they can't even cure!"

This is his favourite theme — good food for good health, bad food for bad health — and Mr. O'Connor is off and running, making connections that sometimes the latest scientific consensus might not, but he does make sense. When he speaks of bees "getting weaker over the past three years because of the B.t. that's everywhere," he is, in his way, bearing witness to something scientists have only recently encountered. Lab observation has shown that bees exposed to B.t. have a higher mortality rate than otherwise; field studies are now attempting to validate the finding.

He has precise observations on birds: "You know what happens with swallows and chickadees? Well they aren't out in the fields on the same dates any more, and they don't eat harmful insects like they used to!" He's also on target when he talks about the butterflies that he calls "papillons monarches" — mixing his French and English: "Well, the caterpillars eat the B.t. pollen on the milkweed leaves, then they die. That's it. No more butterflies."

Here he is backed up by research, and I have a sneaking suspicion he knows it too, though he pretends not to read the specialized farming magazines. John Losey at Cornell University showed in a lab experiment that half the monarch larvae fed on milkweed sprayed with the pollen of B.t. corn died. When *Nature* published the results in 1999, it got a lot of attention, and the biotechnology industry contested it at once. Impossible, they said, because the pollen isn't released at the critical time for larvae growth. It doesn't travel more than an average of two metres, and then they said there was no milkweed in cornfields.

The next year, however, Karen Oberhauser, an ecologist at the University of Minnesota showed all this to be false. Not only *is* the pollen released at that critical time, but milkweed *does* grow in cornfields. What is more, female monarchs prefer laying eggs on milkweed leaves that grow inside cornfields compared to those growing outside.

There is a word that gets Mr. O'Connor's dander up even more than GMOs, and that's *Roundup*. It is a weed killer marketed by multinational Monsanto since 1973. Very popular in North America, it is applied to millions of hectares of crops, particularly corn and soy. Its generic name is *glyphosate*, and it is sold under numerous brands. Unlike most insecticides, it uses a mechanism that bypasses the nervous system and attacks a metabolic path that animals do not possess, thus being far less toxic for them. It also degrades more rapidly in the environment than classic weed killers. Thus it became the favourite tool in eliminating weeds for farmers, agriculturalists, gardeners and those in charge of maintaining highways and rail lines.

Like all farmers, Mr. O'Connor hates weeds much more than insects. "Bastard weeds" he calls them, and it's true enough, since some of the wild plants are capable of a high degree of hybridization. Of course, it's the energy involved in fighting them that most gets to him. If someone tells him he has no choice and ought to just get on with fighting these trespassers, he mutters about how he prepares his soil. "People don't know enough to turn the soil good and deep before sowing," he says, "but I go plenty deep with my plough. Some places I do twice, and when I'm done planting, I walk along and hoe and pull up the weeds by hand. People are in too much of a hurry, and they don't do it right any more."

If you tell him he's doing it the old-fashioned way, taking his sweet time (which is a luxury he can allow himself with so little land to cultivate), he gets warmed up and launches into his standard routine: "What's the point being in a hurry, if you wind up with food that poisons humans and animals, as well as leeching out the land?" Still, Mr. O'Connor, you use weed killers and insecticides like your neighbours, don't you? "Weed killers, yes, if hoeing and plowing aren't enough, but not too much of them, and you've got to rotate your crops every two years, so the weeds don't develop resistance. I vary my crops, and I haven't used chemical insecticides for years. I make a *mélange* of natural products, and only when there's an infestation. I also walk around and check everything, root out insect nests as soon as they show up, and I don't really have any problems as a rule."

Not a word on this mysterious natural *mélange* — he's not going to share his secrets with a city slicker he sees only once in a while when I come to buy vegetables. No point pressing him, but obviously he's been using "organic" methods, especially a multi-faceted plan that involves both controlled, hands-on spreading and population control at the source. So, does this make him an organic farmer? More eyebrow furrowing, shoulder shrugging and sighing: "Doesn't mean much around here. My cereal crops aren't certified organic, and I don't want it. Don't like gimmicks that get you an organic label, specially when nature is taking such knocks from bad farming going on all over the place. No farm is ever insulated, and 'organic' doesn't mean a thing these days unless all the farms in the region are doing it. I told them that at the federation, but they couldn't care less."

That's the verdict once and for all. Mr. O'Connor's fingers play on the scale; then he adds a fistful of carrots to my green beans, offers a grave nod and goes back to work. He's not about to analyze things in any more depth, but he certainly makes sense once his ideas get disentangled from squabbles with the "federation," about which I can't and won't comment. It's true that growing resistance to weed killers and insect killers is the biggest single problem for cereal crops, and a day-to-day nightmare for farmers. As a matter of fact, the continual expansion of industrial crops does make the notion of organic farming a rather misleading one, as well as limiting its impact — at least for now. Mr. O'Connor's view makes sense ecologically too. A holistic approach and good farming practices are needed to avoid the spread of resistance. It is very serious, but the individualism of farmers has so far negated the two famous rules of caution: never use the same pesticide twice in a row, and alternate toxic mechanisms.

I don't know what variety of corn Mr. O'Connor grows, but it's surely one of 262 recognized ones. All of them originate with a plant the ancient

Zea mexicana*, a wild descendant of the primitive teosinte, an ancestor of corn. USDA-NRCS PLANTS Database / Hitchcock, A.S. (rev. A. Chase). 1950.* Manual of the Grasses of the United States. *USDA Miscellaneous Publication No. 200, Washington, DC.*

Mayas called *teosinte*. There's no doubt modern corn comes from the wild plant that existed in southern Mexico about eleven thousand years ago. Archeological and paleontological data tell us it was a thick, low-growing plant with several branches of cobs, not much like the corn we know with just one stalk. The Maya living at high altitudes in southern Mexico domesticated the wild plant by harvesting the grains and planting them, carefully protecting them until harvest time.

Repeating the same experiment year after year, the farmers must have noticed that some plants grew better than others or yielded bigger, tastier cobs; they then selected these seeds to replant or breed with the plants of others, and so on for numerous generations. Artificial selection thus became the beginning of agriculture and led to varieties that were valuable for food in different soils and at different altitudes and latitudes.

In the Americas, the tomato in particular was domesticated, along with the potato and corn, while Europe produced such crops as wheat, barley and rice.

Artificial selection likewise worked in animal husbandry, catering increasingly to humans' food needs, side-by-side with hunting and fishing. Thus, a single species, ours, changed the rules of nature's game and created considerable pressures of selection. Some animal species — like the dodo, the moa and the pie — were completely exterminated, and without realizing it, we have given evolution some very rapid boosts.

We still do, by exterminating the last populations of certain animals or by destroying their habitat, or the systematic hunting or fishing of their best genitors. The mature Newfoundland cod, for example, has evolved to a smaller size in the past sixty years, as a result of massive fishing of the bigger specimens. This is evolution unfolding before our very eyes. Likewise in the plant world: we cultivate fields and forests and fight species deemed undesirable. "The natural" only existed before the arrival of humans. We are the movers and shakers of evolution.

Back to agriculture again: the ability to manipulate genes directly by inserting them into plants and animals might be seen as just another step in artificial selection. Indeed this is the position taken by such regulatory organizations as the FDA (Food and Drug Administration) in the U.S. and the Food Inspection Agency in Canada. To them, inserting a growth gene into a plant or fish is the same as selecting them for the identical characteristics using older methods. Intrinsically, they say, it is no different, and therefore no more dangerous.

This, of course, is hotly debated: environmental groups and some governments, notably in Europe, are opposed to it. Preferring to err on the side of caution, they require a series of safety tests (hence long-term environmental evaluation) before allowing GMOs to be sold. It is an intense debate and amounts to asking if GMOs really are different from artificial selection as practised in traditional agriculture. In what way would the selection carried out by the Maya thousands of years ago be fundamentally foreign to inserting a B.t. gene into the corn raised by Mr. O'Connor's neighbours?

Defenders of GMOs love to say over and over that they are merely producing a desirable trait in a plant or animal (although in an entire population at once) by a precisely focused and limited genetic insertion. It is, they say, simply a rapid, precise and effective method compared with what selectors and nature itself have been doing for hundreds or millions of years. This needs to be looked at from an evolutionary viewpoint.

Rapid? Without doubt. Of course, the insertion of a gene resistant to disease, cold, or drought, for example, can be done quickly, once and for all, and in a controlled fashion in the lab. Precision? Yes, that too seems possible for the same reasons: it is done in the lab with instruments that isolate genes and insert them into organisms whose genomes are known to us. In reality, however, this precision must be subject to caution for several reasons. Often, researchers are not able to insert a gene very, very precisely where it is supposed to go — approximation is acceptable to them; moreover nature itself often uses *several* genes to confer a single trait on an organism, but these are not all employed in the lab. Finally, nature submits this modification in several populations, unlike the quick fix of the laboratory.

Evolution itself uses millions upon millions of trials and errors to do its sorting of large numbers of possible variants. Where one mutation succeeds, many, many others have failed. Finding the combination that will increase the size of a plant or give it more fruit, or lengthen the reproductive season of an animal, natural selection has filtered through numberless genetic combinations. The winner has eclipsed many others and its expression in the organism is therefore solidly established. It is both stable and functional. Going from the many-branched chandelier of teosinte to the single-stalked plant probably took thousands of generations of plants. Of course all this was artificial selection, but it was slow and groping. For an example of natural selection alone, let's look at fish. It took natural selection at least hundreds of thousands of attempts before the halibut developed a protein to help it resist the cold, which has much to do with its survival in certain ocean environments. This means isolated, individual fish in various contexts over hundreds and hundreds of generations.

Artificial selection by modern biotechnology does not do this. It takes shortcuts and does not take time to experiment on large populations in varying situations. It cannot foresee possible failures; in fact, its strategy leaves this out entirely. Since huge amounts of money are invested in it by industrialists, emphasis is placed on positive results and on rapid transfer from lab to field, hence the tendency to skip much-needed and lengthy testing in nature — a dangerous practice.

In light of this, "defective returns" are inevitable. In 1998, it was realized that mustard plants grafted with a gene to resist weed killers could also cross-pollinate with wild plants of the same family — weeds, if you will! This unintended protection of weeds was not good news for the industry, of course. First they denied the problem, then minimized it, only to find themselves admitting everything and cancelling the process. In fact, this flight of genes from GMO plants to wild ones is one of the most worrying ecological setbacks. How can we know the long-term effects of such events? Recent studies lead us to think that, in some cases, we could witness the extinction of wild species and the upheaval of entire ecosystems because of the invasion of these newcomers.

The reality is that we are experimenting on nature in broad sweeps with no way of stopping the machine. Cynics might say that there is nothing new under the sun and this is all déjà-vu, but that is no consolation. When the chemical industry introduced new pesticides thirty to forty years ago, it minimized, even denied, the possible increase in resistance of insects and microbes to these products. Now we are again in the middle of a major environmental crisis. We should have been forewarned.

Mr. O'Connor's fields in the Eastern Townships are a battleground in the fight against GMOs, touted as the irresistible march of progress. Yet there is no scientific answer to the medium- or long-term fallout. His understanding of evolution and of the interaction between species in his environment is fundamentally sound. "I am not against modern technological advances," he says, "even if I don't understand everything they do. We humans have always used poisons to control the natural world, of course, so we can eat, but how do we avoid poisoning the earth, too? We've got to find an answer. This can't just go on."

Avatars of the White Bear

EDDY TAPS HIS FOOT in impatience as he nervously trundles his 440 kilos to and fro along the edge of his pond. He knows his snack is on its way. A short distance behind him, Tiguak, a 240-kilo female, is interested too, though a bit quieter. They are stars at the Aquarium du Québec, two chubby white bears compliantly playing the snack-game before their visitors. Every day, the little 50-kilo trainer perches on a passageway and gets the two giants to do a few pirouettes under the amazed eyes of some children. They stand up on their hind legs, emerge with a splash from beneath the water and generally make waves: with each trick, the artists get some herring, an apple or a large head of lettuce.

Bears in an aquarium? Some biologists actually do consider polar bears a marine mammal, though not all agree. It's true that in the Arctic it's not unusual to come across one of these colossi calmly paddling dozens of kilometres from land. They are capable of swimming hours in a row from one ice sheet to another, and easily dive under it for up to two minutes. Their Latin name leaves no room for doubt about their lifestyle: *Ursus maritimus* (sea bear). In English, we call them polar bears, although they only live around the North Pole and not the South.

In the large-bear family (eight distinct species, plus several sub-species and varieties), the polar bear is the only one with pigment-free fur, although individual colouring may vary toward yellow. It is not an albino, however: the pads on the undersides of its paws and its eyes are coloured with melanin, a natural pigment in all dark-coloured animals. This bear simply doesn't produce them in its fur, rather like an ageing human. Up close, the hairs look as transparent as optic fibre — an effective adaptation to its snow-covered habitat.

It seems, though, that the brown bear (*Ursus arctos*) and the polar bear (*Ursus maritimus*) are very closely related. By the mid-20th century, zookeepers already noticed the two could mate and produce perfectly viable and fertile offspring. In classic terms, this means we are dealing with a single species, although perhaps on the point of splitting into two. This is the famous species barrier so well known to biologists, and which allows them to distinguish between closely related species. Thus the barrier between brown and polar bears is not yet closed. Nevertheless, although their offspring may be fertile, this does not mean they will mate in nature. The division might lie elsewhere: differing reproductive seasons or incompatible behaviour, for instance.

To clear this up once and for all, Finnish paleontologist Björn Kurtén (1924–1988) devoted part of his career to studying bears (the Ursidae family), attempting to establish lineage and ascendancy by comparing the fossils of extinct bears with the bones of contemporary ones. Was he predestined for this work? Björn means "bear" in Swedish. In any event, he did manage, before the advent of modern genetic techniques, to sketch out the evolution of bears summarily but quite accurately.

Kurtén lived and worked at a time when Darwin's theory of evolution was being revised and enlarged to create a *modern evolutionary synthesis*, the theory all modern biologists rely on. In this great process of consolidating knowledge, he is, along with American paleontologist George Gaylord Simpson, at the root of the growing affinity between vertebrate paleontology and the theory of evolution, although his compatriots know him better for the quality of his books popularizing science, or even the works of *paleofiction* (his term) based on his research. Taking constant care to maintain scientific rigour, his stories depict Cro-Magnons living side by side with Neanderthals, although at the time, reliable dating had not yet shown they crossed paths.

Kurtén's work on bears has led to two conclusions. First, *Ursus maritimus* (the polar bear) appeared relatively recently, about 100,000 years ago in the Upper Pleistocene stage. The very first of the genus *Ursus* seem to have emerged in the mid-Pliocene, 3 or 4 million years ago. Characterized by alternating hot and glacial periods, the Pliocene and then the Pleistocene epochs thus saw a rapid adaptive spreading of bears around the planet.

His second discovery was the close link between brown and polar bears. At the time, there was much interest in cave bears, now extinct, which had approximately the same dimensions as modern polar bears, thus making them potential ancestors. From fossils, Kurtén was able to prove that they were completely distinct from one another. The cave bear had a shorter skull with a relatively high forehead, whereas the polar bear has a longer one with a receding forehead. The former had teeth indicating that it was herbivorous, whereas the latter is more of a carnivore.

About forty years after his research, modern genetics tried its hand, and DNA samples from several species of bear were compared by different teams between 1990 and 2000 to determine genetic proximity. The phylogenetic trees thus obtained (see Chapter 2) yielded some commonalities. The Finnish scientist was right: polar bears do indeed appear to be descended from brown bears. Here again, genetic and paleontological data concur.

In that case, is there any way to know exactly which population is involved? Brown bears extend over almost the entire northern hemisphere, and polar bears over the whole of the Arctic, so when did polar bears first make their appearance? A 1996 genetic study by Sandra Talbot and Gerald Fields, both at the Institute of Arctic Biology of the University of Fairbanks, Alaska, offers a rather intriguing answer.

Admiralty, Baranof and Chichagof islands (often called the ABC islands) are in southwestern Alaska, and are inhabited by brown bears, resembling the subspecies *Ursus arctos horribilis*, better known as the grizzly. Although relatively isolated from continental populations, they look like all other Alaskan grizzlies. However, their DNA sequencing is unique with regard to any other on the planet, yet similar to polar bears. They are, in fact, closer to polar bears than any other brown bear. Now, considering the 1,500 kilometres that separated them, how did this happen? Two scenarios were suggested.

Scenario 1: A coastal variant of the brown bear may have lived in northeastern Siberia, then migrated to Alaska 40,000 years ago, giving rise to the polar bear before dying out everywhere but the ABC islands, where they found refuge during the last Ice Age. When it ended about 10,000 years ago, the receding ice allowed brown bears from the south to recolonize Alaska, but not the islands, thus preventing them from sharing genes with the island population. The latter would then be a relic population from 30,000 years previously, hence more closely related to polar bears.

Scenario 2: Since it is mitochondrial genes (tiny units with their own DNA, distinct from the nucleus of the cell in which they find themselves) that underwent study, and these can be transmitted only from a mother to her offspring, it may be that at some point a polar female coupled with a brown male on the island and thus gave birth, injecting her own mitochondria into the island population. Only a fresh study of the nucleus' DNA itself could resolve this issue. Amazingly enough, no one has yet undertaken it.

Whatever its precise origins, one cannot help admiring how far the polar bear has gone in adapting to its environment — the true lord of the iceberg!

White as snow for camouflage, its fur also provides excellent insulation, so much so that the bear will suffocate at above 10°C. Along with a thick layer of body fat, this insulation allows it to avoid hypothermia underwater, as well as from glacial winds leading to temperatures as low as minus 70°C. Like all bears, of course, it stores up reserves of fat in late spring and late autumn, and yet remains active in all seasons; only pregnant females take refuge in a lair when November rolls around, and they stay there until they have given birth, surviving on reserves and lowered body temperature.

Their paws, broad and round at the extremities, are virtual paddles when it comes to swimming for hours. For running on the ice, the paws have another advantage: lots of soft little bubbles which serve as suction cups that help keep them from slipping and sliding.

Despite their imposing mass and seeming nonchalance, this plantigrade is capable of impressive speed, running at up to forty kilometres per hour for hundreds of metres. Humans are understandably afraid of a predator capable of such speeds, and the Inuit speak freely of it in their hunting tales. Nanuk, as the bear is known, is the most feared and respected of animals, and the occasional non-native venturing out onto the ice is also well warned of the danger. A researcher explained to me recently that the Inuit guides hired to protect scientific missions on the ice fear more than anything else the prospect of a hungry polar bear emerging from the fog at full speed, lured by the smell of another mammal. For this reason, they are always armed and ready to fire. The moment a bear is spotted nearby and appears to be approaching, they fire warning shots to scare him off. "It doesn't always work," the researcher adds. "Sometimes we have to give up on taking a measure or a sample and beat it out of there."

Seals are at the top of polar bear menus. This equal-opportunity carnivore will also dine on walruses, fish and the carcasses of beached whales, although seals are definitely its staple. It finds them with its sharp sense of smell and eyesight, often tracking them in the vicinity of an air hole in the ice and waiting patiently for one to surface. (Contrary to legend, it has never been known to cover its muzzle with one paw when on the hunt.)

Temperatures throughout the North in the past fifty years have risen three to four degrees Celsius, and they continue to rise faster still. In the past thirty years, a million square kilometres of summer ice has disappeared, which is twice the size of France (only one-tenth the size of Canada, but still a lot). The problem is, of course, universal and worldwide, manifested here in regional terms. Our behaviour as southern, energy-gobbling consumers is responsible for serious global warming with major implications for ecosystems a long way off.

The Intergovernmental Panel on Climate Change (IPCC) foresees a world temperature rise of 1.8 to 4°C in temperatures and 18 to 59 cm in sea levels. A nordic country like Canada could undergo even greater variations, particularly in the northern and western portions of the country. The federal government predicts a dire future for the summer sea ice, expecting it to shrink by 50 to 100 percent. Since the 1980s, ice in the region of Churchill on Hudson Bay in northern Manitoba has been melting three weeks sooner than previously. A fairly southerly location for polar bears, the region harbours a population of close to 4,500. As spring draws to a close, this warming-induced melting of the ice forces these "southern" bears to cease seal hunting before they have built up the reserves of fat they need. Inversely, on the shores of the bay in October, although already weakened by their summer diet (fish, algae, wild berries, bird eggs, caribou carcasses), they must wait longer for ice to form to be able migrate north and resume seal hunting.

The consequences are already quite visible, and have made the polar bear a symbolic martyr for the struggle against climate change. Around Churchill, bear safaris are organized in specially designed vehicles rolling over the frozen tundra, where tourists can see and photograph them, but the reality is rather sad, for if they were not starving, they wouldn't be there. These are de-natured specimens, hanging around the edge of Churchill so they can raid the garbage dumps. Their health is becoming worse and worse, and so is their reproduction rate. Biologists have confirmed the weight-loss of individuals in the southernmost zones.

Their behaviour, too, is affected, as suggested by two disturbing studies published in spring 2006. The first, by Alaska researchers, mentions drowning deaths among polar bears after swimming too long in the Beaufort Sea along the North Coast. Starving bears, in search of the sea ice that melts too soon in spring and freezes too late in the fall, swim for days before finally drowning of exhaustion. This type of incident, never observed before 2004, has become increasingly frequent. The second study, published jointly by Canadian and American researchers, reports three confirmed cases of predation and cannibalism among polar bears in only three months' observation, all in the same sector of Alaska. In thirty-four years of polar bear studies, we have only just begun to see hungry bears eating one another.

Industrial pollution seeping into their food is far from helping things. At the top of their food chain, the bears gradually accumulate as part of their fat the chemicals spilled into the water: bi-phenyl derivatives (BPCs for instance) and heavy metals. These lead to toxicity levels that usually cause premature death or increasingly frequent foetal malformation in newborns. In 2006, the International Union for the Conservation of Nature added the polar bear to its endangered species list for these very reasons. Latest estimates foresee their numbers decreasing by 30 percent in the next forty-five years as a result of global warming. If the summer ice disappears completely, the remaining 24,000 (15,000 of them in Canada) may become completely extinct.

The polar bear is confined to its habitat by a highly specialized diet. Unlike brown or black bears, which are far more general in their needs, the polar bear's diet is distinct and precise, and its ecological niche very narrow. Although carnivorous and opportunistic like other bears, it depends essentially on one prey, and one prey only, seals. This has held true for at least 10,000 years. The shrinking coastal sea ice and pack ice have special importance for this animal, for as they melt, we must ask whether the bears will adapt and follow the seals wherever they go. Or, will they instead alter their diet and perhaps feed more on fish? The Arctic does not provide many choices. The polar bear will also have to deal with oil spills, which have destroyed more and more habitat, as well as the organochloric residue building up in its body fat and internal organs.

Its future is anything but rosy, to say the least. Even science, which can predict a 50 percent reduction in summer sea ice within a century, cannot foretell the outcome for this species. From a long-term perspective one might argue that the disappearance of the polar bear will not be such a catastrophe for the planet, or even the arctic ecosystem. The story of life, after all, is one in which millions of species disappear, many of them from climate change.

Yet there is an important difference worth considering. This time, we are the ones responsible for climate change. Our actions, our consumption of fossil fuel and natural resources have caused this warming, this melting of sea ice and this threat to the species. Of course, to be fair, one can say that global warming has some natural causes too, but uppermost is our knowledge that human activity is making an extensive contribution.

So, can we save the polar bear? There is no guarantee, but we can try. The Arctic is vast, and spectacular conservation efforts are still possible. The introduction of strict hunting restrictions since 1973 has already been a step in the right direction. The bears need extensive natural reserves where they can be fully protected. Since it has been possible to save at least some lions and elephants in southern Africa, it would seem reasonable to try and do the same with polar bears, which have enormous tourist, heritage and cultural value.

Remarking that today bear viewing is creating more revenue than hunting, and that tourism in protected areas has become a major industry, Matthias Breiter, a German-born author now living in central Canada, summarizes the situation in a way I find at once subtle and profound:

> Bears have fascinated mankind for millennia. They captivate us today as [they] did our ancestors. The good of a bear can be measured in many ways. Yes, a dollar value can be attached to its existence. But while its spiritual and cultural value is much less tangible, it is by no means less significant. In North America, we are fortunate to share our country still with these awe-inspiring creatures. Grizzlies, black bears and polar bears have all lost part of their former ranges. Where bears still roam, a fragile truce often prevails with man and beast. Usually it is the people rather than the bear that lack the ability to adapt.[27]

The choice is up to us, and despite some pessimistic viewpoints, this is one case where it is not too late. If we do nothing, of course, the only survivors we will be able to admire in the next century will be the captive descendants of the well-known pair at the Aquarium du Québec, Eddy and Tiguak.

Sign of the Loon

IT IS NOW NEARLY three years since I began writing this book, and very little has changed at the foot of the elephant mountain. Life goes on with its myriad interactions in the plants and the small animal community. What is three years in the life of an ecosystem? Less than a lightning bolt in the night of time. If humans don't commit too much violence to it, life just goes on. To understand more easily how that works, two things are all we need: natural selection (Darwin's stroke of genius) and what can be called "the play of genes" — the bundle of ideas delivered up to us by genetics, embryology and the most up-to-date molecular biology.

Fortunately, our section of the lake isn't seriously threatened by home building. It's not as though developers and individual buyers aren't interested; it's just that there's nothing left for sale. Most of the surrounding woods are privately owned, and people generally don't want to sell. In fact, municipal regulations forbid any new development or drastic changes along the banks (and the owners' association sees to this, especially preventing the cutting of trees). The large bog on the way out is protected, and can never be dried out, or, we hope, built on. It's our natural reservoir, a refuge for our birds, our frogs, our muskrats, our red foxes and our beautiful plants. There is a home development project on the neighbouring lake, close enough for us to hear the carpenters' saws and heavy machinery from time to time, but we treat it as something far from us. Of course, it's an illusion; the natural integrity of the place is losing ground a bit at a time, just as it is everywhere else. At least we're putting up a fight like the plucky Gauls in the time of Asterix, and maybe we'll actually get to keep a bit of it. Who said resistance was futile?

Oscar our squirrel died last year. He never showed up in spring, and a few months ago, another red squirrel appeared instead, smaller, just as busy, but not nearly as noisy. And so one individual gives way to another by death or competition. Darwin told us this was fiercest among members of the same species; thus we call the new tenant Oscar Bis (Encore).

The lake is stable, with water levels varying by scarcely a few centimetres each year, with the same fish and the same otter wreaking havoc among them. So discreet is this otter that we've never actually seen her, although her tracks are evident in the snow.

Last fall brought lots of rain and wind storms, with strong gusts that knocked down a number of trees. We had to cut down one of the large pines ourselves after it died. It provided us with a huge load of firewood for the fireplace. Inside, the logs are stacked two metres high, and the pine smell is wonderfully fresh.

This bit of southern Quebec forest is like hundreds of other ecosystems. There is no great revolutionary upheaval to be found, but life is in constant motion. Darwin described this continuity spiked with constant happenings very well in his last paragraph of *On the Origin of Species*:

> It is interesting to contemplate an entangled bank, clothed with many plants of many kinds, with birds singing on the bushes, with various insects

> flitting about, and with worms crawling through the damp earth, and to reflect that these elaborately constructed forms, so different from each other, and dependent on each other in so complex a manner, have all been produced by laws acting around us. These laws, taken in the largest sense, being Growth with Reproduction; Inheritance which is almost implied by reproduction; Variability from the indirect and direct action of the external conditions of life, and from use and disuse; a Ratio of Increase so high as to lead to a Struggle for Life, and as a consequence to Natural Selection, entailing Divergence of Character and the Extinction of less-improved forms. Thus, from the war of nature, from famine and death, the most exalted object which we are capable of conceiving, namely, the production of the higher animals, directly follows.[27]

He concludes his *magnum opus* a few lines later with the splendid and oft-quoted words:

> There is grandeur in this view of life, with its several powers, having been originally breathed into a few forms or into one; and that, whilst this planet has gone cycling on according to the fixed law of gravity, from so simple a beginning endless forms most beautiful and most wonderful have been, and are being, evolved.[28]

Far from our bit of the lake, out in the wide world, the theory of evolution continues to be contested. It's not something the media pay a lot of attention to, nor scientists for that matter. There is no scandal about it, just an air of déjà vu and perhaps an occasional skirmish: here a school opting to teach creationism as an alternative to neo-Darwinism, there a politician confessing he doesn't believe in the theory of evolution. If protest goes largely unnoticed, it's because it occurs essentially in the private sphere, or in very discreet areas of public discourse: grass-roots school and social organizations. Some observers consider it the outgrowth of the strengthening right-wing conservatism currently in power in North America. Now in the majority and listened to by elected politicians, these citizens and voters are expressing themselves more and more at the local level.

It's important to remember that in North America, it is state education departments and provincial ministries that control educational curricula, and local school boards also have their own decisions to make in the matter. If one of them decides to suppress Darwin's ideas, or teachers decide not to touch on them in class, the harm is done, and both science and education take a step backward.

As a matter of fact, four states in recent years (Alabama, New Mexico, Nebraska and Kansas) have removed the theory of evolution from high school exam questions, although Kansas reversed itself in 2001. Numerous school boards, like Dover County, Pennsylvania, have introduced the teaching of creationism as an alternative to the theory of evolution, and many private schools affiliated with fundamentalist churches, do as they wish, ignoring official programs. In Quebec, some such schools, it was learned in the autumn of 2006, are quite simply illegal and "off the grid."

Thus creationism is taught as truth in lower grades, and the theory of evolution as a "hypothetical" alternative in science classes. Teachers hired by religious communities spontaneously adopt this revisionist approach, and if not, the parents — wielding great influence in these schools — bring pressure to bear and demand that education reflect their values. When we add the growing influence of sects like the Jehovah's Witnesses, we find ourselves in a boiling cauldron of anti-Darwinism. Since the papers hardly mention this, the phenomenon is poorly documented and rarely researched, and consequently we have cause to fear that we are educating a generation diverted from science.

I am fully aware that the opinions I give here are personal ones and that the threat to science teaching needs deeper study. Permit me simply to express my disquiet based on impressions or anecdotes which may not be alarming in themselves. I would like to conclude this little personal distress call with two real-life incidents.

Four years ago, I was invited to give a presentation in two pre-university institutions in Quebec, one a CEGEP (a public junior college), and the other a private college, to students mostly between the ages of eighteen and twenty-one. The talk was on the human genome, and I wanted to gauge their knowledge with a few questions, including: "Do you think that humans and chimps have the same genetic code or different ones?" In the first case, I was dealing with students in the social sciences, and in the second with students in pure and applied science. I was struck by the fact that almost all the students in the second group answered correctly, while only a third of the first group did. What, then, had happened to their high school biology?

More recently, I wrote in *Québec Science* on sex determination in the platypus. This animal, emblematic of Australian fauna, is certainly strange enough, looking as it does, like a huge mole with webbed feet, a beaver-like tail, a beak like a duck — not least, it lays eggs! Yet, for all this, it has mammary glands. It is one of four monotremes, the only mammals that lay eggs instead of giving birth to living foetuses.

To my surprise, this publication landed me in a mini-debate on creationism that I was not at all expecting. One indignant reader made it clear he was not at all impressed by my article: "Once again," he wrote, "evolutionists are incapable of explaining the origin of platypi, merely because it doesn't fit their theory." Another pointed out that the platypus possessed "20 physiological features inexplicable in a mammal," thus providing "the grain of sand that collapses the theory of evolution."

A third sent a more courteous letter, ending thus: "Despite all its tricks, modern biology cannot explain the complex sexual system of platypi. I cannot wait for a new synthesis to emerge that will reconcile scientific knowledge with the revelation of the Creation of the world."

To my first correspondents I replied with references to scientific articles that answered their objections (although I seriously doubted that anything could shake their certainty). Fossils have in fact been found dating back to the Mesozoic. On this basis, it is possible to explain why the platypus exists only on the Australian continent and is classified among a group appearing very early in mammalian history, having diverged from other mammals (such as marsupials) about 210 million years ago. As to its physiological "incongruities," they fade away when seen in comparison with a broad range of animals.

To the third correspondent, I replied with due respect for his opinion, explaining that I didn't think science and faith could find a common language on this point, being on totally different planes. If a believer in religion accepts that species appear and evolve through natural selection, fine, dialogue is possible (this is in fact the position of numerous scientists who are also believers). However, if this person asserts that God, not a natural process, created species and does as He wills with them, then it is not. The two positions are irreconcilable. Either it is a fact that species follow one another, emerging from other species and disappearing in a changing and uncontrolled environment, or they appear by magic at the hands of a creator whimsical enough, it must be said, to create millions of differing forms of life and then eliminate them by the millions.

Certainly the modern theory of evolution does not explain everything, but if life still holds many mysteries, at least science has patiently built up the more convincing file on them with a solid theory that can be tested and verified. And verified it has been. From this perspective, the modest platypus is by no means incomprehensible, nor is it a weak link in Darwin's chain of reasoning — bizarre perhaps, but in no way inexplicable.

It is April, and the lake is slowly melting, the land surrounding the cottage wet and muddy. Everywhere, we can hear the water lapping. Sheets of snow still hang on under the trees, looking pitifully fragile, as does the skin of ice on the lake, which has begun to sink bit by bit, first around the edges, then refreezing a little at night. The nearby roads have bulges in the asphalt, and the dirt roads are now free of snow. Smells, of course, are part of spring's rebirth: today it smells of wet earth, and already the first green shoots are visible — the first plants are on the way, with the promise of their shapes and colours.

First a good mid-April snowstorm — you have to expect them here — then the weather is mild. A few more days, and the thermometer regularly consorts with temperatures above zero. This is the time for plants, animals and humans to come out and play at the foot of the elephant mountain. Stretched out on the dock, with heavy parkas and cushions beneath our heads, we make the most of the late afternoon sun. We feel like plants turning to face the heat and light, perhaps even trying for a little photosynthesis ourselves. Oh how we've missed that solar heat! Only in the nordic countries does one greet it with such feverishness. (I mentioned this to some Finns one day, and they too described themselves as sun-crazy in the first days of spring).

Listen to the silence of the forest. Not complete silence, just a series of little wind-sounds, and a rippling of thawing water from the shore over to the right. High up, a breeze shakes the pine and cedar tops; it's a long, gentle breath, not like winter winds. I recall going outside on winter mornings after a snowfall, and there was the cold, the smell of cold and the sound of feet crunching in fresh snow. What was striking then was the wind blowing through the tops of the cedars and pines, and the immense straight and slender trunks reaching to the sky. I thought of the solitude of those great trees and of their quiet strength throughout the cold. Looking at them seemed to give me heart. These giants lived their lives slowly, but they were imperturbable. Now here they are, already full of life, shaken by this light breeze, sap running in their trunks and branches. At the top, needles are already growing, and cones will soon be out, first tiny, then visible and full to bursting.

The return of Oscar Bis is one sign that can be relied on. His schedule never fails. This winter, he was constantly in and out of his nest, often for short spurts when the sun was at its highest and it wasn't too cold, a little ball of fur tearing across the snow. Now he's out more often, and letting out his "chir-t-t-t." Each day, he's busier.

There is an incident that happened a few weeks ago in early March that keeps returning to me. I recall evening falling and the shadows lengthening around the cottage. All is calm. It is a magical hour I love so well. On the edge of the lake in my parka, I don't feel the cold. A few snowflakes hang in the air, when suddenly behind me, there's a cracking of branches, and I spin round, thinking my neighbour has fallen at the end of my property, with his enormous polar boots on and ready for a chat perhaps, and a "Hey neighbour, you there?" It's not him though. It's not André. It's a huge brown horse stuck in a snowdrift and struggling toward me! Then, in a tenth of a second, it metamorphoses into a moose, huge and headed straight for me! It can't be. He hasn't spotted me yet, so I move, I clap my hands, and then he stops abruptly about twenty-five metres away, between two pine trees.

He raises that enormous head, and I see his long, arched muzzle and two black eyes staring at me. Then he lurches heavily off to the right on those long legs of his, seemingly so fragile in the thick blanket of snow, off down the bank, and in a few bounds there he is. Soon he's upright on the ice, standing slowly and sniffing the air, gently placing one huge hoof after another with surprising elegance for a 500-kilogram creature. He trots lightly and easily across the frozen lake and through the thin dusting of snow from the previous days.

What a sight, and what a feeling to see him so close. How could that happen? A moose (*Alces alces*) has weak eyesight but makes up for it with acute hearing and sense of smell; that is why I'm surprised he didn't notice me earlier. He was coming down the slope quite quickly, taking the shortest route straight ahead. I've called him "Edgar égaré" (Wandering Edgar) since then, for he's part of a family that live on the north hill, I believe, and he was far from home. Normally they won't venture this far and don't cross the lake.

The picture haunts and never ceases to amaze me. First of all, this animal has devoted his energy and talent to the task of survival, then to reproduction: both are imperative. Second, he is a member of a present-day species, one little twig on the tree of life, living like all of us, the limited existence of a tiny twig, with evolution just continuing on its way. Every species on that tree, past and present, even the most remote, like the moose or the little-known scud, are connected. To be exact, they all have a last common ancestor from which their respective histories then diverged. All this diversification happened in about 3 billion years from primitive life forms, "from such a simple beginning," as Darwin would say.

We know all this today, the better to study the real world and push back the frontiers of ignorance, while appreciating the world around us. By offering a deeper look at ourselves, at our existence, at our place in the universe, science leads us to a vision which is profoundly poetic. It evokes images, calls up emotions, adds meaning to the world. And what tremendous luck we have. So rare and unique is our privilege, that by developing our brains and our consciousness — all this the fruit of evolution — we are animals capable of understanding our existence and even the world itself across time.

All at once, in the sky above us, there's a call from somewhere around the elephant mountain, a sort of sharp barking. A turn of the head, and there is a cloud of black dots moving through the sky. Wild or Canada geese are approaching, as always, on their way north. Their distinct bi-syllabic cries are now clearly audible, "a-honk, a-honk." Maybe they'll stop in a nearby swamp for a day or two, long enough to feed and build up their strength, or maybe they'll just keep on going right past us to the next meeting point, around Sorel on the edge of the Saint Lawrence, or even further, near Cap Tourmente close to Quebec City.

There's the squadron now, flying in a superb "V" formation, cries resonating like a familiar and comforting symphony. Few have hailed their return with the eloquence of Aldo Leopold, the American ecologist, in *A Sand County Almanac*:

> They weave low over the marshes and meadows, greeting each newly melted puddle and pool. Finally, after a few *pro forma* circlings of our marsh, they set wing and glide silently to the pond, black landing-gear lowered and rumps white against the far hill. Once touching the water, our newly arrived guests set up a honking and splashing that shakes the last thought of winter out of the brittle cattails. Our geese are home again![30]

One has to admire wild geese and celebrate their return as the surest sign of nature's strength and endurance, a strength both unerring, yet changeable across many generations.

Yet, while I hail this segment from the immense diversity of living things, here I am, thinking about *our* loons too. Will they be back this year? It is late April, too soon to tell, but I look forward to their return and their magnificent plaintive wail. Three weeks later, and I have my answer.

One evening, from a neighbouring lake comes a powerful call, the unmistakeable cry of the loon (*Gavia immer*). There it is again, and again many times: three long plaintive notes, perhaps the male's first call upon his return. These large aquatic birds are often faithful to one location, and I'd like to think that it's the same one as last year. Perhaps I'm kidding myself, but still there are a few clues: this one, like last year's, doesn't much mind human presence, since he lives on the other lake and feeds on ours. These lakes are smaller than the ones usually mentioned in scientific publications describing their habitat. They are also shallower and more frequented by humans, but the bay in which I suspect they nest is quiet enough, without roads or houses nearby.

A few days after these night cries, I saw him fishing one morning in the middle of our lake, diving and reappearing regularly a few moments later, still half-submerged, his long beak horizontal. A few more days and a fresh surprise, though much awaited — there are two of them. Again, I wonder: are these two a new couple, or the same ones as last year? Now there they are, plain as day, fishing and patrolling the lake from the middle, attentive, half-submerged. Then a tip over and, rumps in the air, gone again for thirty to forty seconds for another fruitful dive.

I can imagine how fast they move under the water, the surprised fish caught in those sharp beaks. *Gavia immer* is one of the Gaviidae, a group of birds appearing for the first time in the Eocene period — although this is disputed, since neither fossils nor phylogenetic analysis has settled the question — and of which only five species remain: this common loon, the red-throated loon, the Arctic loon, the Pacific loon and the white-billed diver. The common loon is the most widespread and the most southerly. In Quebec alone, their population is around 50,000 — not a species threatened with extinction. Its habitat, though, is shrinking.

Popular wisdom has it that "if the lake has a loon, the water is clear and clean." This, in fact, is true and can be explained two ways. First, the loon is a visual hunter, and depends on water clarity to detect its prey. Second, it is a greedy fish eater, consuming up to 250 kilos of fish in one summer. Therefore, it must feed in fish-rich lakes like ours. Our lake certainly is well stocked with fish, perhaps because modern vacationers are not the inveterate fishermen their forefathers were. Traditions do die out! Little matter, if it allows us a privileged encounter with such a magnificent bird, a friend from the very depths of time.

There is no doubt that the loon's cry was heard by the first humans to walk this spot of earth about 10,000 years ago when the ice retreated, and hearing its cry, they must have felt the same emotions as we do now.

The sign of the loon is the sign of our intimate connection to all living things, in this case a bird whose ancestor first appeared over 45 million years ago. It is a sign of life's deepest unity, the loon taking its place beside all the species that have co-existed for millions of years: all of them cousins, evolving together in a constantly changing environment, and connected by the same eternal thread of DNA. Finally, for us modern humans, it's the sign of an ecological balance that must be maintained, for loons at the lake mean they can reproduce, even in the presence of humans. If modern science can help us understand that, maybe it can also help us protect these riches.

Calendar of Life

THIS CALENDAR IS inseparable from the Earth's geological calendar. Geological history as documented from rocks begins in the Archean Eon about 4 billion years ago. About half a billion years later, the first life forms appeared. During the major part of the Precambrian Eon (Proterozoic) between 2.5 billion and 545 million years ago, life developed very slowly.

The Primary or Paleozoic era began about 545 million years ago with the explosion of life in the Cambrian Period.

The Mesozoic or Secondary era (Trias, Jurassic and Cretaceous periods), notable for the presence of dinosaurs and the first mammals, began approximately 250 million years ago.

The Cenozoic Era (Tertiary and Quaternary periods) began 65 million years ago: mammals and especially primates diversified, and towards the end hominids appeared.

The Quaternary Period (Pleistocene, then Holocene epochs) began 1.8 million years ago, with cycles of glaciation and de-glaciation.

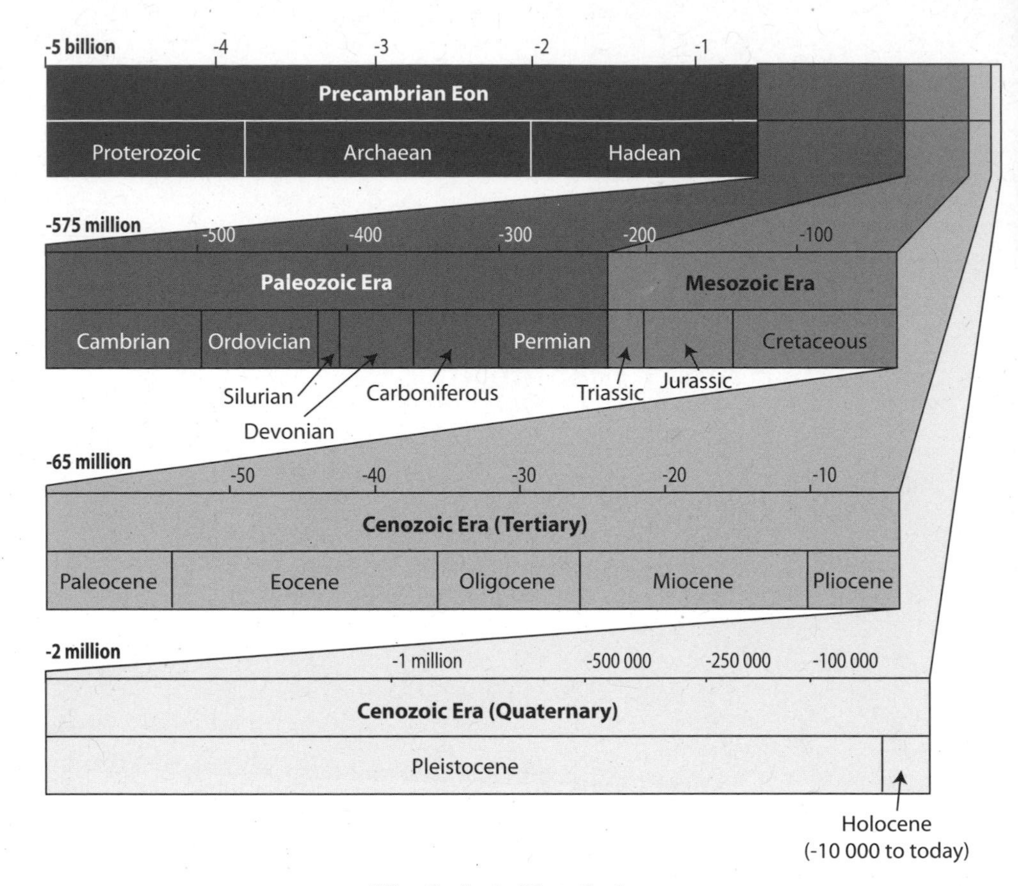

The Geologic Time Scale

Glossary

■ *Acanthostega*
A tetrapod living 363 million years ago and discovered in Greenland in 1987. Possibly the very first tetrapod.

■ Artiodactyls
Animals with an even number of toes (e.g. pigs, cows, camels and hippopotami), as opposed to perissodactyls, with uneven numbers of toes (e.g. horses, rhinoceri).

■ Base
Elementary chemical component of which there are four: A (adenine), T (thymine), C (cytosine) and G (guanine). These are always in pairs of facing opposites on the rungs of the DNA ladder: A opposite T, C opposite G. (In RNA, U — for uracil — replaces T). The sequence of bases constitutes a genetic "word," or gene.

- *Basilosaurus*
  The small-legged, presumed ancestor of whales, discovered in Louisiana in 1832. It lived 37 to 40 million years ago.

- Biodiversity
  The variety and variability of all living organisms. This includes, according to the International Union for Conservation of Nature, "genetic variability within species and their populations, the diversity of associated complex species and their interactions, and those of the ecological processes they influence or put into effect (ecosystemic diversity)."

- Cetaceans
  Order of marine mammals including whales (with teeth or baleens), dolphins and porpoises.

- Clade
  Evolutionary unit including all descendants of an ancestral species. All members share at least one characteristic derived from that ancestor.

- Cladistic
  Method which bases hierarchy and lineage of living creatures on the distribution of states derived from common characteristics.

- Co-evolution
  Evolution as it concerns two or more species, such as a parasite and host, during which changes in one species influence the evolution of the other(s).

- Deep Time
  Expression of geological time popularized by John McPhee. Over 3.5 billion years, from the Primary era to the Quaternary, life developed, first slowly, then more rapidly.

- DNA
  Initials for deoxyribonucleic acid, the molecule that maintains heredity, consisting of a linear string of nucleotides. Its structure is composed of two threads twisted in a double-helix, the bases of which form the rungs of a spiralled ladder.

- Ecosystem

  A collection of organisms living in a given milieu, such as a forest, and physical environmental elements with which they interact (soil, climate).

- *Eusthenopteron*

  A large fish about 1 metre long, living 375 million years ago, and found in Miguasha, Quebec.

- Evo-devo

  Stands for embryological *evolution* and *development*, a new scientific discipline at the junction of evolution and embryology, with a good measure of genetics. Evo-devo concerns itself in particularly with genetic and developmental comparisons between species.

- Exaptation

  A feature that undertakes a new function during evolution, such as the panda's thumb. It has replaced the term "pre-adaptation."

- Gene

  A specific segment of DNA containing coded information permitting a cell to manufacture proteins.

- Genetic code

  A system of correspondence among triplets of bases in DNA and amino acids (elementary units of protein structure).

- Genome

  Set of genetic instructions in a given organism, carried by genes.

- Homeotic or hox genes

  Genes governing the construction of axes and members in an animal. Hox genes are the homeotic genes in vertebrates.

- Homology/analogy

  Structures are homologous if constructed on the same model of embryonic development, whether or not they fulfill the same function (e.g. the forelimbs of moles, whales, horses, bats and humans). On the other hand, "analogous" describes a similarity in function between structures of different origin (e.g. insect wings and bird wings).

■ Invertebrate

Animal without a vertebral column. Invertebrates make up most of the animal kingdom, including arthropods, molluscs, echinodermata, annelids. They do not form a clade as vertebrates do.

■ Molecular Clock

Dating technique based on the hypothesis that, for a given genome subset and corresponding proteins, the number of mutations per millions of years is constant.

■ Mutation

Change in the structure of a gene or chromosome. Some can be triggered (e.g. by radiation), but most occur spontaneously in the course of duplication of genetic material from one generation to another.

■ Nucleotide

Basic unit of DNA, composed of sugar-phosphates bonded to one of the four bases, A, C, T, G.

■ Odontocetes

Whales with teeth, including dolphins, porpoises, narwhals, belugas, orcas or killer whales, sperm whales, black pilot whales and common beaked whales (all others are baleen whales).

■ *Pakicetus*

The most ancient and primitive known ancestor of whales. Lived 49 million years ago.

■ Phylogenetics

Study of the formation and evolution of organisms with a view to establishing their relationships.

■ Selection

Process resulting from the superior survival and reproduction of certain individuals due to their genetic makeup (genotype). Selection may be natural, as demonstrated by Darwin, or artificial, as exercised by humans in the domestication of certain species.

■ Sequencing

Process for determining the order of amino acids in proteins, or of nucleotides in segments of corresponding DNA.

■ Species

A collection of similar individuals sharing hereditary physiological and morphological characteristics, and capable of reproducing together.

■ Taxonomy

Science of the laws of classification. Arranging living things according to kingdom, phylum (formerly "branch"), class, order, family, genus and species.

■ Tetrapods

Important group of land vertebrates, including aquatic animals such as whales, descended from a common finned ancestor living 375 million years ago.

■ Ungulates

Hooved mammals, including artiodactyls, perissodactyls and proboscidea (elephants).

■ Vertebrates

Animals with backbones. A branch of living creatures, including fish, amphibians, birds, reptiles and mammals.

Notes

1 Stephen Jay Gould, *The Panda's Thumb: More Reflections in Natural History* (New York: W.W. Norton, 1980), 12.

2 Charles Darwin, *On the Origin of Species by Means of Natural Selection, 1st ed.* (London: John Murray, 1859), 48.

3. Cyrille Barrette, *Le miroir du monde, évolution par sélection naturelle et mystère de la nature humaine* (Sainte-Foy, P.Q., Éditions MultiMondes, 2000), 91.

4. Frère Marie-Victorin, *Flore Laurentienne*, (Montréal: Presses de l'université de Montréal, 1964), 243.

5 Charles Darwin, *On the Origin of Species*, 180.

6 Ibid., 181.

7 Ibid., 181.

8 Ibid., 182.

9 Richard Dawkins, *The Blind Watchmaker* (London: Longman, 1986), 272.

10 Ibid., 274.

11 Paul Hebert, Mark Stoeckle, "Barcode of Life," in *Scientific American* (October 2008), 88.

12 William Flower, "On Whales, Past and Present, and Their Probable Origins," *Notices of the Proceedings of the Royal Institution of Great Britain*, 10 (1883): 375.

13 Jean-Henri Fabre, *Souvenirs entomologiques: Études sur l'instinct et les m?urs des insects*, vol. 1 (Paris: Robert Laffont/Bouquins, 2004), 233.

14 Sean Carroll, *Endless Forms Most Beautiful: The New Science of Evo-Devo* (New York: W.W. Norton, 2005), 208.

15 Ibid., 209.

16 François Jacob, *Le Jeu des possible: Essai sur la diversité du vivant* (Paris: Fayard, 1977).

17 Ibid., 66.

18 Charles Darwin, unpublished notebooks.

19 David Tyler, "Natural Selection in Galapagos Finches," in *Biblical Creation Society Magazine*, (1994).

20 Stephen Jay Gould, *The Panda's Thumb: More Reflections in Natural History* (New York, W.W. Norton, 1980), 3.

21 Stéphane Peigné, quoted in *Journal du CNRS* (March 2006): 5.

22 *Charles Darwin, The Descent of Man and Selection in Relation to Sex* (London: John Murray, 1871), 153.

23 Robert Sapolsky, "The 2% Difference" in *Discover* (April 2006): 45.

24 Stephen Palumbi, *The Evolution Explosion: How Humans Cause Rapid Evolution Change* (New York: W.W. Norton, 2002), 130.

25 Ibid.

26 Salmon taxonomy is complicated by both the common usage of the word "salmon" and by commercial categories associated with what we associate with salmon. What we commonly call salmon is a subfamily (Salmoninae) of the great salmon family Salmonidae which has undergone a complex pan-arctic evolutionary radiation. The Salmonidae consists of three sub-families: Coregoninae (Whitefish, 78 species), Thymallinae (Graylings, 12 species), and Salmoninae (Salmon and trout, 102 species). Worldwide there are 7 genuses. Three genuses of Salmoninae are found in North America: *Salmo, Oncorhynchus and Salvelinus*. Oncorhynchus has a Pacific distribution and is limited to 14 species. *Salmo* (29 species) and *Salvelinus*

(49 species) have a broader distribution generally east of the Rockies. These three genuses include all the fish known popularly as salmon, trout and chars. See also David R. Montgomery's article "Coevolution of the Pacific Salmon and Pacific Rim Topography," in volume 28, number 12, of the journal *Geology* (2000): 1107–10.

27 Matthias Breiter: *Bears: A Year in the Life* (Richmond Hill, Canada: Firefly Books, 2005), 13.

28 Charles Darwin, *On the Origin of Species*, 548.

29 Ibid.

30 Aldo Leopold, *A Sand County Almanac with Essays on Conservation from Round River* (New York: Oxford University Press, 1949), 20.

Bibliography

AUTHOR'S NOTE: Biographies of Charles Darwin and accounts of the theory of evolution are numerous, and the bicentenary of his birth has brought a wealth of new books and reprints. This post-anniversary work is a personal selection and in no way claims to be exhaustive, but it does cover numerous aspects of the contemporary discussion of Darwinism. Furthermore, the sources cited here are above all intended for a broad reading public and do not include the most specialized works. It has focused on works originally published in French or English, primarily with regard to relevance, rather than choice of language. Finally, numerous scientific articles I have relied on are not included, although the interested reader may easily find them using the listings given in various chapters (principal author, year of publication, university affiliation and/or title).

Arnould, Jacques, Pierre-Henri Gouyon, and Jean-Pierre Henry. *Les Avatars du gene: La théorie néodarwinienne de l'évolution*, Paris: Belin,1997.
Barrette, Cyrille. *Le miroir du monde, évolution par sélection naturelle et mystère de la nature humaine*. Sainte-Foy, P.Q., Éditions MultiMondes, 2000.
Breiter, Matthias: *Bears: A Year in the Life*. Richmond Hill, Canada: Firefly Books, 2005.
Browne, Janet. *Charles Darwin. Voyaging*. London: Pimlico, 1996.
——. *Charles Darwin. The Power of Place*. London: Pimlico, 2003.
Carroll, Sean B. *Endless Forms Most Beautiful: The New Science of Evo-Devo*. New York: W.W. Norton, 2005.
——. *Remarkable Creatures: Epic Adventures in the Search for the Origin of Species*. New York Houghton, Mifflin, Harcourt, 2009.
Combes, Claude. *Les associations du vivant: L'art d'être parasite*. Paris: Flammarion, 2001.
Coynes, Jerry. *Why Evolution is True*. New York: Viking, 2009.
Crick, Francis. *Life Itself: Its Origin and Nature*. London: Macdonald, 1981.
Darwin, Charles. *Journal of Researches into the Natural History and Geology of the Countries Visited During the Voyage of HMS* Beagle *Around the World*. London: John Murray, 1845. (Darwin's true best-seller, this is easy-to-read and better known in subsequent editions under the shorter title *The Voyage of the Beagle*. He was not gifted for brief titles.)
——. *On the Origin of Species by Means of Natural Selection*. 1st ed. London: John Murray, 1859; http://darwin-online.org.uk/ or http://www.talkorigins.org/faqs/origin.html/.
——. *The Descent of Man and Selection in Relation to Sex*. 2 vols. London: John Murray, 1871.
——. Unpublished notebooks.
Darwin, Francis, ed. *Charles Darwin's Autobiography, with His Notes and Letters Depicting the Growth of* The Origin of Species. New York: Henry Schuman, 1950.
Dawkins, Richard. *The Blind Watchmaker*. London: Longman, 1986.
——. *Climbing Mount Improbable*. London: Viking, 1966.
——. *The Selfish Gene*. Oxford: Oxford University Press, 30th anniversary edition, 2006.
——. *The Greatest Show on Earth: The Evidence for Evolution*. London: Free Press, 2009.
Dennett, Daniel. *Darwin's Dangerous Idea*. New York: Simon and Schuster, 1996.
Desmond, Adrian, and James Moore. *Darwin's Sacred Cause: How a Hatred of Slavery Shaped Darwin's Views on Human Evolution*. New York: Houghton Mifflin Harcourt, 2009. (A fresh perspective on Darwin's beliefs by two of his biographers.)
Eldredge, Niles. *Darwin: Discovering the Tree of Life*. New York & London: W.W. Norton, 2005.
Fabre, Jean-Henri. *Souvenirs entomologiques: Études sur l'instinct et les m?urs des insectes, vol. 1 et vol. 2*, Paris: Robert Laffont/Bouquins, 2004.
Gould, Stephen J. *Ever Since Darwin: Reflections in Natural History*. London: Burnett Books, 1978.

——. *The Panda's Thumb: More Reflections in Natural History*. New York: W.W. Norton, 1980.
——. *Hen's Teeth and Horse's Toes*. New York: W.W. Norton, 1983.
——. *Wonderful Life: The Burgess Shale and the Nature of History*. New York: W. W. Norton, 1989.
Hebert, Paul, and Mark Stoeckle. "Barcode of Life," in *Scientific American* (October 2008).
Hall, Brian, ed. *Fins into Limbs: Evolution, Development, and Transformation*. Chicago: University of Chicago Press, 2007.
Jacob, François. *Le Jeu des possible: essai sur la diversité du vivant*. Paris: Fayard, 1977.
——. *La souris, la mouche et l'homme*, Paris: Odile Jacob, 1997.
Keynes, Randal. *Annie's Box: Charles Darwin, His Daughter and Human Evolution*. London: Fourth Estate, 2001.
Leopold, Aldo. *A Sand County Almanac with Essays on Conservation from Round River*. New York: Oxford University Press, 1949.
Marie-Victorin. *Flore laurentienne*. Montreal: Presses de l'université de Montréal, 1964.
Maynard-Smith, John. *The Theory of Evolution*. 3rd ed. Cambridge: Cambridge University Press, 2008.
Miller, Kenneth. *Only a Theory: Evolution and the Battle for America's Soul*. New York: Viking, 2008. (My favourite critique of creationism.)
Monod, Jacques. *Le hasard et la nécessité: Essai sur la philosophie naturelle de la biologie moderne*. Paris: Seuil, 1970.
Montgomery, David R. "Coevolution of the Pacific Salmon and Pacific Rim Topography." *Geology* 28, 12 (2000).
Palumbi, Stephen. *The Evolution Explosion: How Humans Cause Rapid Evolution Change*. New York: W.W. Norton and Company, 2002.
Prothero, Donald. *Evolution: What the Fossils Say and Why it Matters*. New York, Columbia University Press, 2007.
Quammen, David. *The Song of the Dodo: Island Biogeography in an Age of Extinctions*. London: Hutchinson, 1996.
Ruse, Michael. *Darwin and Design: Does Evolution Have a Purpose?* Cambridge, Mass: Harvard University Press, 2003.
Sapolsky, Robert. "The 2% Difference" in *Discover* (April 2006).
Scott, E.C. *Evolution vs. Creationism: An Introduction*. Westport, Conn: Greenwood, 2004.
Shubin, Neil. *Your Inner Fish: A Journey into the 3.5 Billion-Year History of the Human Body*. London: Allen Lane, 2008.
Suzuki, David, and Wayne Grady. *Tree: A Life Story*. Vancouver: Greystone Books, 2004.
Tort, Patrick. *L'effet Darwin: Sélection naturelle et naissance de la civilisation*. Paris: Le Seuil, 2008.
Weiner, Jonathan. *The Beak of the Finch: A Story of Evolution in Our Time*. London, Jonathan Cape, 1994.
Wilson, Edward O. *The Diversity of Life*. Cambridge, Mass.: Harvard University Press, 1992.

Zimmer, Carl. *Evolution: The Triumph of an Idea*. New York: HarperCollins, 2001.
——. *At the Water's Edge: Fish with Fingers, Whales with Legs, and How Life Came Ashore but Then Went Back to Sea*. New York: Touchstone Books, 1999.

About the Author & Translator

JEAN-PIERRE ROGEL is a reporter for *Découverte*, a popular science program on Radio-Canada, the French-language counterpart to the CBC. He also writes for *Québec Science* magazine. He is a passionate naturalist and has written extensively about how and where one can find the evidence for evolution and how everything on earth is connected in a web of being. Born and educated in France, he immigrated to Canada in the mid-1970s and has been living in Montreal ever since. This is his fourth book.

NIGEL SPENCER is one of Canada's outstanding translators and has won Governor General's Awards for two consecutive translations of Marie-Claire Blais' novels *Thunder and Light* and *Augustino and the Choir of Destruction*. For Ronsdale Press he has also translated Marie-Claire Blais' collection of short stories *The Exile & The Sacred Travellers* and her collection of plays *Wintersleep*. He makes his home in Montreal.

Index

[*Boldface numbers refer to pictures*]

Abzhanov, Arkhat, 70
Acanthostega, **75**, 75–81
Alberch, Father, 78–80
Ambulocetus, 46
Aquarium du Québec, 125, 132
Aristotle, 42
artificial selection: in bacteria and viruses, 106–7; breeding and, 16–17, 120–22; environmental impact of, 102–3, 111–14, 119–20, 129–32; GMOs and, 117–19, 122–24

bacteria: antibiotics and, 106–7; *Bacillus thuringiensis* (B.t.), 117–19; co-evolution in, 103–7; *E. coli*, 30; *Staphylococcus aureus*, 106; Streptococcus A, 106
Barcode of Life Initiative, 34–38
Barrette, Cyrille, 18–19
Basilosaurus, 42, 45–46
Bateson, William, 56
bear: brown (*Ursus arctos*), 126–28; evolution of, 126–27; grizzly (*Ursus arctos horribilis*), 128; Inuit and, 129; polar (*Ursus maritimus*), 125–32
Béland, Pierre, 40
biodiversity: Barcode of Life Initiative, 34–38; human impact on, 40, 50–51, 102–3, 106–7, 111–14, 117–24, 129–32; origin of, 18

Boisserie, Jean-Renaud, 46
Breiter, Matthias, 132
butterfly: Apollo (*Parnassius apollo*), 60–61, **61**; buckeye (*Junonia coenia*), 61–63; GMOs and, 118–19; false eyes of, 60–64

Carroll, Sean B.: *Endless Forms Most Beautiful*, 61–63
chimpanzee (*Pan troglodytes*), 90–98
Clack, Jennifer A., 75–79
Coates, Michael, 76–80
Combes, Claude: *Les associations du vivant*, 103–4
convergent evolution, 29, 86–87
corn: GMO, 117–19; teosinte, 120–21, **121**
creationism: intelligent design and, 49, 72, 91–92; origin of, 15; public opinion on, 2–3, 49; versus evolution, 71–73, 135–38
Crick, Francis, 18

Daniel, Thomas, 32–33
Darwin, Charles: on common lineage, 25; *The Descent of Man and Selection in Relation to Sex*, 90–91; finches and, 67–69; humans and, 89–92; impact of theory by, 1–3; on natural selection, 15–17, 134–35; *On the Origin of Species*, 1–2, 15–16, 22–23, 42, 89, 92, 134–35; on speciation, 114; on tree of life, 22–25, 29–30; on whales, 42
Davis, Dwight, 87–88
Dawkins, Richard: *The Blind Watchmaker*, 27–29; on Darwin, 2
deWaard, Jeremy, 38
DNA: definition of, 26–27; discovery of, 18; human, 93–94; mutation of, 56–59; sequencing, 28–29, 94; species classification using, 28–29, 34–38, 44–46, 113, 127–28
Dobzhansky, Theodosius, 2, 18
dolphin, 49–51
Dorudon, 45–46

Eldredge, Niles, 102, 114
embryology: anomalies and, 56–58; evo-devo and, 70, 101; human evolution and, 97; polydactyly and, 78–80, **79**
Eusthenopteron, **76**, 76–78, 80–81
evo-devo: butterflies and, 61–64; definition of, 3; finches and, 70–71; humans and, 96–98; regulator (homeotic) genes and, 56–59, **58**, 61–64, 80, 87–88, 96–98
extinction: human impact on, 102–3, 122, 124, 130–31; natural selection and, 22, 72, 135

Fabre, Jean-Henri, 60
Fields, Gerald, 128
finch: *Carpodacus mexicanus*, 69; Darwin and, 67; *Geospiza fortis*, 67, **68**, 69–72; *Geospiza magnirostris*, 67, **68**, 69–70; other varieties of, 67–68; #5110, 71; *Tiaris obscura*, 69
Fisher, Ronald, 18
fixism, 15
Flower, William, 42–44
fossils: *Acanthostega*, **75**, 75–81; *Ambulocetus*, 46; *Basilosaurus*, 42, 45–46; *Dorudon*, 45–46; *Eosalmo driftwoodensis*, 110; *Eusthenopteron*, **76**, 76–78, 80–81; *Homo habilis*, 95; *Ichthyostega*, 78; *Pakicetus*, **45**, 45–46; *Rodhocetus*, 46; *Simocyon batalleri*, 85–86; *Tulerpeton*, 78

Fraser River, 109–12
fruit fly (Drosophila): butterfly evolution and, 61–62; heredity of, 17; mutations in, 56–59, **58**

Galapagos: Daphne Major, 68–70, 71; Darwin's studies in, 16, 67; Genovesa, 68; Grants' studies in, 66–73
Gatesy, John, 44
Gehring, Walter, 57–59
genes: Barcode of Life Initiative and, 34–38; of butterflies, 61–64; DNA sequencing and, 28–29, 94; of finches, 70; of fruit flies, 56–59, **58**; genome data banks and, 29; heredity and, 17–18; of humans, 95–98; regulator (homeotic), 57–59, **58**, 61–64, 80, 87–88, 96
genes, types of: *aniridia*, 58–59; *antennapedia*, 57; *bithorax*, 57; BMP4, 70; *calmodulin*, 70; cytochrome b, 35; cytochrome c oxydase I (COI), 35; *distal-less*, 59, 63; *engrailed*, 63; *eyeless*, 58–59; FOXP2, 96; MYH16, 95; PAX6, 59; *small eye*, 58–59; *spalt*, 63; *tinman*, 59
genetics: development and, 3, 56–59, 61–64, 70, 80, 87–88; genetic code and, 5, 18, 26–27; GMOs and, 117–19, 122–24; human, 93–96; population, 17–18, 25–26; species classification and, 28–29, 34–38, 44–46, 113, 127–28; tetraploid, 110
geologic time scale: Archean Eon, 145; Cenozoic Era, 145; Devonian Period, 74–75; diagram of, **146**; Eocene Epoch, 110; Mesozoic Era, 137, 145; Miocene Epoch, 110; Paleozoic Era, 145; Pliocene Epoch, 110, 127; Pleistocene Epoch, 127, 145; Precambrian Eon, 145; Proterozoic Eon, 145; Quaternary Period, 145; Upper Pleistocene stage, 127
Gesner, Abraham, 81
Gingerich, Philip, 44–46
global warming, 111, 129–32
GMOs: *Bacillus thuringiensis* (B.t.) and, 117–19; ecological impact of, 118–19, 122–24
Goldschmidt, Richard, 56
goose, Canada (*Branta canadensis*), 141
Gould, Stephen Jay: *The Panda's Thumb*, 5, 83–86; theory of punctuated equilibrium and, 114
Grant, Peter & Rosemary, 65–73
Great Chain of Being, 24
Greenland, 74–75

Hebert, Paul, 34–38
Hendry, Andrew, 113
Herschel, John, 11, 16
hippopotamus (*Hippopotamus amphibius*), 43–44, 46, 48
human (*Homo sapiens*): evolution of, 49, 90–98; *Homo habilis*, 95; impact on biodiversity, 40, 50–51, 102–3, 106–7, 111–14, 117–24, 129–32; taxonomy of, 33
Huxley, Thomas, 90

Ichthyostega, 78
Intelligent Design, 49, 72, 91–92
Intergovernmental Panel on Climate Change, 130
International Union for Conservation of Nature, 102, 131
International Whaling Commission, 50

Jacob, François, 64, 85, 101
Jardin botanique de Montréal, 21–25

Kurtén, Björn, 126–27

Leopold, Aldo: *A Sand County Almanac*, 141
Linnaeus, Carl, 14, 33
loon, common (*Gavia immer*): 12–13, 141–43
Losey, John, 118

macroevolution, 72–73
Marie-Victorin, Brother: *Flore Laurentienne*, 19
Maynard-Smith, John, 23
Mayr, Ernst, 18
Mendel, Gregor Johann, 17
microevolution, 72–73
Miguasha, 76, 81
molecular clock, 27–28
molecular phylogenics, 29–31, 44
Monet, Jean-Baptiste de: *Zoological Philosophy*, 15
Mont Ventoux, 59–60
Monteverde Cloud Forest, 32–33
Montgomery, David, 111
moose (*Alces alces*), 139–40
Morgan, Thomas Hunt, 17
mutation: anomalies and, 56; bacteria and viruses and, 106–7; evolution theory and, 18, 72, 101, 114; GMOs and, 117–19, 122–24; molecular clock and, 27–28; speciation and, 71–73, 95–96; regulator genes and, 56–59, 63–64; tetraploids and, 110–11

natural selection: arguments against, 72–73, 137–38; Darwin on, 15–17, 134–35; François Jacob on, 64; human influence on, 16–17, 102–3, 106–7, 111–14, 117–24, 129–32; in parasites and hosts, 104–7; in pitcher plants, 20; theories of, 2, 16–18, 101
neo-Darwinism, 18

Oberhauser, Karen, 118–19
On the Origin of Species: debate over, 89, 92; natural selection discussed in, 16, 134–35; publication of, 1–2; tree of life discussed in, 22–23; whales discussed in, 42
orchid, 85

Pakicetus, **45,** 45–46
paleontology: hominids and, 90–91; panda and, **84**, 85–86; platypus and, 137; polar bear and, 126–27; salmon and, 110; tetrapods and, 75–82; theory of punctuated equilibrium and, 114; whales and, 42–47
Paley, William, 91, 92
Palumbi, Stephen: *The Evolution Explosion*, 107
panda: giant (*Ailuropoda melanoleuca*), 83–85, **84**, 86–88; *Simocyon batalleri*, 85–86; small or red (*Ailurus fulgens*), 86–88
parasitism: co-evolution and, 103–7; obligatory parasites and, 106; "Red Queen Hypothesis" and, 105; simple, 103; small liver fluke and, 104; symbiosis, 103
Peigné, Stéphane, 86
Perrin, Jean, 70
phylogenetics: of bears, 126–28; DNA sequencing and, 28–29, 94; of humans, 90–98; molecular clock and, 27–28; molecular phylogenics,

29–31; of panda, 85–88; of platypus, 137; of salmon, 110–15; of tetrapods, 75–82; of whales, 42–46
pine, white (*Pinus strobus*), 13–14
pitcher plant, purple (*Sarracenia purpurea*), 19–20
platypus (*Ornithorhynchus anatinus*), 137
proteins: DNA sequencing and, 28–29, 94; genetic code and, 26; GMO corn and, 117; halibut and, 123; human, 94–96; homeotic box and, 57; molecular clock and, 27–28
proteins, types of: ASPM, 96; cytochrome c, 28; myosin heavy chain 16, 95

reproduction: of finches, 68–69, 71, 73; of orchids, 85; of parasites, 103–4; of peas, 17; of platypus, 137; of polar bears, 130; of sockeye salmon, 111–13; of viruses, 107; of whales, 40–41; of white pines, 13–14
Ricket, W. E., 112
Rodhocetus, 46
Royal Society for the Advancement of Science, 89–90
Russell, Donald, 45

Saguenay River, 40
salamander, axolotl (*Ambystoma mexicanum*), 63–64
Salesa, Manuel, 86
salmon: Atlantic (*Salmo salar*), 108–11; chinook (*Oncorhynchus tshawytscha*), 114; *Eosalmo driftwoodensis*, 110; origin of, 110–11; pink (*Oncorhynchus gorbuscha*), 112; and rapid evolution, 112–15; sockeye (*Oncorhynchus nerka*), 109–12; tetraploid, 110–11
Sapolsky, Robert, 96–98
Shubin, Neil, 78–80
Simocyon batalleri, 85–86
Simpson, George Gaylord, 18, 127
speciation: in bears, 126; in finches, 71–73; gradualism, 17, 114; in humans, 90–98; in salmon, 111–14; theory of punctuated equilibrium and, 114
spontaneous generation, 15
squirrel, American red (*Tamiasciurus hudsonicus*), 9, 11, 134
St. Lawrence River, 39–42, 47
Stedman, Hansell, 95
Stoeckle, Mark, 37–38
synthetic theory of evolution, 18, 114

Tabin, Clifford, 70
Talbot, Sandra, 128
taxonomy: Barcode of Life Initiative and, 34–38; DNA sequencing and, 28–29; of humans, 33; Linnaeus' system and, 14, 24–25, 33–34; molecular phylogenics and, 29–31; of pandas, 86; of whales, 42–46
tetrapod: *Acanthostega*, **75**, 75–81; *Eusthenopteron*, **76**, 76–78, 80–81; *Ichthyostega*, 78; *Tulerpeton*, 78
transformism, 15
tree of life, 22–25, 29–30, 140
Tulerpeton, 78

universities: Alberta, 110; Arizona, 44; British Columbia, 65, 113; Cambridge, 75; Chicago, 76, 105; Cornell, 118; Fairbanks, 128; Guelph, 34–38; Harvard, 78; McGill, 66, 113; Michigan, 44; Minnesota, 118–19; Oxford, 2, 27, 89–90; Pennsylvania, 95;

Rockefeller, 37; Stanford, 96; Washington, 113; Washington State at Seattle, 111; Windsor, 113; Wisconsin-Madison, 61

Van Valen, Leigh, 105–6
virus: co-evolution in, 103–7; HIV, 107

Wallace, Alfred Russell, 1–2
Washington Zoo, 83–84
Watson, James, 18
whale: beluga (*Delphinapterus leucas*), 39–42, **41**; blue, 39; hunting of, 40, 50; evolution of, 42–48; rorqual, 39–40, 47
Wilberforce, Archbishop Samuel, 90
woodpecker, hairy (*Picoides villosus*), 11–12